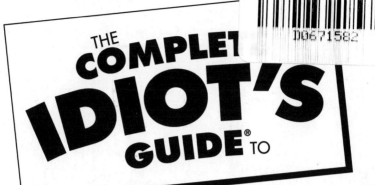

THE

COMPLETE
IDIOT'S
GUIDE® TO

Skype
for PCs

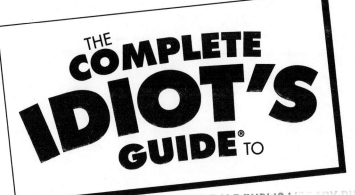

THE COMPLETE IDIOT'S GUIDE® TO

Skype
for PCs

by Andrew Sheppard

ALPHA

A member of Penguin Group (USA) Inc.

To my loving wife, Susan, and wonderful son, Nathaniel.

ALPHA BOOKS

Published by the Penguin Group

Penguin Group (USA) Inc., 375 Hudson Street, New York, New York 10014, U.S.A.

Penguin Group (Canada), 10 Alcorn Avenue, Toronto, Ontario, Canada M4V 3B2 (a division of Pearson Penguin Canada Inc.)

Penguin Books Ltd, 80 Strand, London WC2R 0RL, England

Penguin Ireland, 25 St Stephen's Green, Dublin 2, Ireland (a division of Penguin Books Ltd)

Penguin Group (Australia), 250 Camberwell Road, Camberwell, Victoria 3124, Australia (a division of Pearson Australia Group Pty Ltd)

Penguin Books India Pvt Ltd, 11 Community Centre, Panchsheel Park, New Delhi—110 017, India

Penguin Group (NZ), cnr Airborne and Rosedale Roads, Albany, Auckland 1310, New Zealand (a division of Pearson New Zealand Ltd)

Penguin Books (South Africa) (Pty) Ltd, 24 Sturdee Avenue, Rosebank, Johannesburg 2196, South Africa

Penguin Books Ltd, Registered Offices: 80 Strand, London WC2R 0RL, England

International Standard Book Number: 1-59257-551-X
Library of Congress Catalog Card Number: 2006925724

08 07 06   8 7 6 5 4 3 2 1

Interpretation of the printing code: The rightmost number of the first series of numbers is the year of the book's printing; the rightmost number of the second series of numbers is the number of the book's printing. For example, a printing code of 06-1 shows that the first printing occurred in 2006.

Printed in the United States of America

Note: This publication contains the opinions and ideas of its author. It is intended to provide helpful and informative material on the subject matter covered. It is sold with the understanding that the author and publisher are not engaged in rendering professional services in the book. If the reader requires personal assistance or advice, a competent professional should be consulted.

The author and publisher specifically disclaim any responsibility for any liability, loss, or risk, personal or otherwise, which is incurred as a consequence, directly or indirectly, of the use and application of any of the contents of this book.

Most Alpha books are available at special quantity discounts for bulk purchases for sales promotions, premiums, fund-raising, or educational use. Special books, or book excerpts, can also be created to fit specific needs.

For details, write: Special Markets, Alpha Books, 375 Hudson Street, New York, NY 10014.

Publisher: *Marie Butler-Knight*
Editorial Director: *Mike Sanders*
Managing Editor: *Billy Fields*
Acquisitions Editor: *Tom Stevens*
Development Editor: *Jennifer Moore*
Production Editor: *Megan Douglass*

Copy Editor: *Michael Dietsch*
Cover Designer: *Bill Thomas*
Book Designers: *Trina Wurst/Kurt Owens*
Indexer: *Angie Bess*
Layout: *Chad Dressler*
Proofreader: *Aaron Black*

Contents at a Glance

Appendixes

Contents

Foreword

Congratulations, you're hooked on Skype!

I'm still dragging people into Skypeland, chatting it up to the annoyance of many taxi drivers. It's easy, because Skyping with friends is fun. And Skyping at work saves so much time and effort, especially if you work with others.

Free calls. Chat, conferencing, SMS, voice, and video all in one nice simple secure package. It can change your life as much as e-mail and the web have.

I want you to suck Skype's marrow, all the juicy bits. This book gets you running. Do you know the stuff that's embarrassingly obvious once someone points it out? All in here, blush free.

There's more, of course. Andrew Sheppard and the *Complete Idiot's Guide* series would never leave you just inside the front door; they show you around. This book is full of tips and tricks that turn you into a power Skyper. Some will save you money, many will save you time, and they'll all do that other important thing: cultivate your relationships.

Skype is for conversation. The heartfelt, the pizza order, the tech support line, the Mandarin tutor. Moments that build friendships, link our families, and get things done. Skype helps piano teachers, relief workers in distant disaster zones, and parents with kids at college. Even the occasional call from strangers—particularly from overseas—can be fun and educational.

Cheap and convenient aside, it comes down to Skypenomics. We're in an era of overwhelming distraction, where time is dear and we want the most from every minute. Skype helps you focus on the people in your life. Productivity where it matters.

Skype helps us work better. Skype Journal is a virtual company, with authors spread over many countries. We use Skype, blogs, and wikis to do our work. Because Skype is social software, it's easy for us to quickly form teams around new projects, discuss breaking news, troubleshoot problems, and just keep ourselves from being lonely in our home offices. Because of Skype, we get more done in less time, freeing us to teach and consult. All without dropping the ball.

This book's solid grounding will leave you prepared as Skypenomics evolve into buying and selling information, entertainment, education, services, and other intangibles.

Read on, Skype on, and hook your friends.

Phil Wolff

Phil Wolff is Editor in Chief of Skype Journal (http://SkypeJournal. com), a daily news magazine covering Skype and the New Conversation for users, suits, and geeks. He leads Skype Journal Consulting's strategy and IT practice and is an alumnus of Adecco SA, Compaq Computer, LSI Logic, Bechtel National, Wang Labs, and the Naval Supply Systems Command.

Introduction

When the ground on which we stand moves, it is the result of a tremor, an earthquake, or a tectonic shift. Internet telephony started as a tremor only a few short years ago. It is now an earthquake. And within a decade from now it will have resulted in a tectonic shift in how phone calls are made the world over. Indeed, it will radically alter how people communicate in all manner of ways, not just by voice. Clearly, the future of telephony is the Internet, for which geographic location and distance don't matter.

Skype is one of the key players in shaking up how phone calls are made and, equally importantly, how much they cost. Skype and other Internet-telephony providers are giving consumers new ways to phone and new ways to save money. New ways to phone because Internet telephony is based on new technologies that open up new possibilities in terms of what you can do with a phone call. New ways to save money because, in addition to making geographic location and distance largely irrelevant, Internet telephony enables phone calls to be made in ways and at a cost limited only by your ingenuity.

This book is about new ways to phone and new ways to save money using Skype. Other Internet-telephony providers are mentioned only when they offer a better deal than Skype or provide some functionality not currently provided by Skype.

Whether you are just starting out with Skype, or whether you've been using Skype for a while, you will find in this book fresh ideas on how to get more out of Skype.

How This Book Is Organized

The book is organized into six parts.

Part 1, "Up and Running with Skype," helps you install and configure Skype with the minimum of fuss. It shows you the key elements of the Skype software program that runs on your PC, how to navigate its graphical user interface, and how to carry out common tasks, such as making a call.

Part 2, "Advanced Skype and Skype Services," introduces some of the more advanced features of Skype, such as face-to-face video calls. It also describes in detail each of Skype's fee-based services. Skype comes with a lot of free features, but Skype's functionality can be further extended by subscribing to one or more of Skype's fee-based services. Part 2 is rounded out by a chapter that describes how Skype can save you money when making calls, and what you can do to maximize these savings.

Part 3, "Customize and Extend Skype," shows you how you can configure Skype to better meet your individual needs. That is, you can bend Skype to your way of doing things, and this section shows you how.

Part 4, "Privacy and Security," describes the problems and pitfalls that can beset any program that connects to the Internet. Potential privacy and security threats are described, and remedies are suggested. With the knowledge gained from this section, you can give yourself the visibility you desire within the Skype community, and you can block many of the threats that can harm your PC.

Part 5, "Troubleshoot Skype," helps you when things go wrong. Like any other piece of software, Skype can act up at times, and this section offers tools and tips for diagnosing and fixing problems. Also, Skype does not support 911 emergency services, 411 directory services, or regular fax machines, but I show you workarounds for these problems so that you can reclaim much of this lost functionality when switching to Skype.

Part 6, "Accessibility and Usability," is aimed at making Skype more accessible and usable for people with and without disabilities. We all want Skype to be easy to use and to work in harmony with our way of doing things. For some, particularly people with disabilities, this is a necessity, while for others it makes using Skype more enjoyable. Part 6 includes tips and tricks to improve your Skype experience.

Extras

Throughout the book, terminology, interesting facts, warnings, and tips are highlighted in sidebars that expand your knowledge and understanding of Skype, but without spoiling the flow of the book. Four sidebar types are used:

Something to Try

Here you'll find tips on using Skype's features, shortcuts that simplify common tasks, or simply fun ways to use Skype.

def•i•ni•tion

These sidebars introduce terminology that might be unfamiliar to you.

Caution!

Heed these warnings to avoid pitfalls and problems when using Skype.

Something Worth Knowing

The information included in these boxes helps you get the most out of Skype.

A Few Notes

Skype supports Windows 2000 and Windows XP. However, to avoid repetition in the text, all menu navigation is described for Windows XP. The differences between Windows XP and Windows 2000 are minor. Where you see the instruction **Start: All Programs** in Windows XP, it is **Start: Programs** in Windows 2000. Similarly, **Start: Control Panel** in Windows XP is **Start: Settings: Control Panel** in Windows 2000.

This book was written and tested with the latest version of Skype available at the time of writing: 2.0.0.103.

Acknowledgments

Tom Stevens, my acquisitions editor at Alpha Books, saved my life—albeit metaphorically—more than once during the writing of this book! As a sounding board for ideas, and when a path needed to be cleared through contractual and administrative issues, Tom was there. Thanks Tom. To the many other people at Alpha Books who helped make this book a reality, I extend to them my heartfelt thanks. Although too numerous to thank individually by name, I do feel that special thanks and recognition are due my development editor, Jennifer Moore, who did a wonderful job with the book.

In addition to the thorough formal technical review of the book carried out by Lawrence Hudson, the book was informally—though equally thoroughly, as is his style—reviewed by Adam Harris. Better known as "Gladiator" on the Skype user forums, Adam is the number one contributor (by a wide margin) to Skype's own community forums, and I extend my thanks to everyone in that community. Adam brought a rigor and clarity to the review process that can only come from a pre-eminent expert on Skype, for which I thank him.

Special Thanks to the Technical Reviewer

The Complete Idiot's Guide to Skype for PCs was reviewed by an expert who double-checked the accuracy of what you'll learn here, to help us ensure that this book gives you everything you need to know about Skype. Special thanks are extended to Lawrence Hudson.

Trademarks

All terms mentioned in this book that are known to be or are suspected of being trademarks or service marks have been appropriately capitalized. Alpha Books and Penguin Group (USA) Inc. cannot attest to the accuracy of this information. Use of a term in this book should not be regarded as affecting the validity of any trademark or service mark.

Part 1

Up and Running with Skype

If you're new to Skype and want to get it up and running with minimum fuss, this is where you should start. Using Skype is so simple that in no time you'll have made your first Skype call!

The chapters in this part of the book provide an overview of Internet telephony in general and Skype in particular, and then proceed to show you how to download, install, and configure Skype for your computer. You'll also be given a brief tutorial on Skype's graphical user interface.

About Skype and What It Can Do for You

In This Chapter

- ◆ Finding out what Skype can do for you
- ◆ Understanding how Skype works
- ◆ Taking a tour of Skype's features
- ◆ Looking at alternatives to Skype

Skype is a free computer program you can use to make telephone calls over the Internet. In addition, Skype offers a fee-based subscription service that allows you to add a variety of bells and whistles to the free basic calling feature.

Skype is a softphone, or a program that emulates the functions of a telephone. However, Skype does far more than simply make voice calls. You can also use it to make conference calls and video calls, to chat, and to transfer files. And thanks to the advantages of its underlying technologies, Skype can do many other things for you that a traditional telephone simply cannot do and which I will cover in detail later in this book.

What Skype Can Do: An Overview

Skype-to-Skype calls are free, as are video calls, chat, and file transfers. So even without becoming a fee-paying Skype subscriber, you can do an awful lot for free!

Skype's fee-based subscription services enhance Skype so that you can also make calls to, as well as receive calls from, regular and mobile phones. For additional fees, you can add on voicemail, custom ring-tones, and other optional services.

Skype's phenomenal success, in large part, is due to the rich—and growing—feature set you get simply by installing Skype on your computer. Also, Skype has received a good deal of praise for the simplicity of its user interface and its ease of use, which have done much to contribute to its success. The Skype softphone runs in a small window on your PC, with a graphical interface that is surprisingly easy to navigate given the number of features it supports, as the following figure illustrates.

The Skype software client.

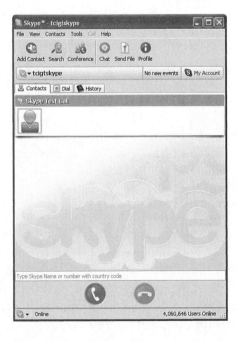

As mentioned previously, you can augment Skype's basic feature set by subscribing to any of the following fee-based services:

- ◆ **SkypeOut:** Make calls to regular and mobile phones.

- ◆ **SkypeIn:** Receive calls from regular and mobile phones.

- ◆ **Voicemail:** Record voice messages from callers.

- ◆ **Skype Zones:** Use all of Skype's services while traveling.

- ◆ **Personalise Skype:** Add custom ringtones and pictures to the Skype softphone.

- ◆ **Skype Control Panel:** Manage Skype subscription services for a group of Skype users (mostly of interest to business users of Skype).

I explain these services in detail later in this book. For now, all you need to know is that you can pick and choose the features and services you want. In contrast to many regular and mobile phone plans, which are typically quite inflexible about what features are included with your service plan, with Skype you only have to pay for those features you want. Plus, Skype doesn't require you to sign any complex or long-term contracts. That's right: you can cancel anytime—often receiving a refund for unused services (see Chapter 9). Try that with your phone company or mobile carrier!

A Brief History of Skype

Skype is not the first *P2P* software to be made widely available via the Internet and to achieve mainstream status. Indeed, the people who founded Skype, Niklas Zennström and Janus Friis, are the same people who created the P2P file sharing program Kazaa, which is the most downloaded (at 363 million as of September 2004) software in the history of the Internet. Skype clearly has something of a pedigree.

def•i•ni•tion

P2P stands for "peer-to-peer," which is a type of network configuration in which all the computers are interconnected in such a way that they are treated equally; that is, they are all peers. Skype's network is P2P.

Skype was founded in 2002, but the domain names Skype.com and Skype.net weren't registered until April 2003. August 2003 saw the release of the first public beta, or trial, version of the Skype softphone (often also referred to as the Skype client). In June 2004, Skype introduced its first fee-based service, SkypeOut, which enables Skype users to call regular phone numbers all over the world at very low rates (as low as $0.021 per minute). SkypeIn, which is a fee-based service that allows people to call Skype users from regular phones, debuted in March 2005. Skype continues to add new features and services at a rapid pace.

Something Worth Knowing

In October 2005, Skype was acquired by eBay, which also owns PayPal. eBay's goal is to use Skype to enable its buyers and sellers to communicate more effectively and so reduce the "friction" inherent in buying and selling things. Overall, this is good news for Skype and for Skype users. Not only will eBay's user base be encouraged to use Skype—significantly adding to the number of people already using the program—it gives Skype the backing of an Internet powerhouse that has lots of marketing and financial muscle.

What's Next for Skype?

Apart from continued improvements in the basic technologies that underpin Skype and the improvements in quality they bring with them, what's next for Skype? For one thing, new Skype-enabled hardware and Skype-compatible hardware is being introduced. For example, Skype-enabled phones and handhelds are being sold that already have Skype installed, configured, and ready to go right out-of-the-box. We can also count on Skype continuing to add new features and services. But perhaps the biggest impact of Skype will be felt once it becomes truly mobile, in a "connect from anywhere and at any time" sense. Already, some cities

def•i•ni•tion

WiFi is short for "wireless fidelity" and is a wireless network technology that has a limited range that is typically measured in the tens of feet indoors and the hundreds of feet outdoors. Some cities have begun providing free citywide WiFi coverage by placing WiFi transceivers in public places. More cities are sure to follow.

in the United States are offering free city-wide *WiFi* Internet access. For many people, this means that a handheld or other portable wireless device running Skype is a close substitute for a mobile phone; the only difference is that calls on Skype are free (to other Skype users) or at very low rates (for calls to regular and mobile phones)!

How Skype Works

Skype is very different from most traditional Internet applications, in which a client (that's your PC running a piece of software) connects over the Internet to a server dedicated to the task of—you guessed it—serving clients.

Skype's P2P Architecture

The Skype network can be thought of as a mesh of computers that connect with each other over the Internet. As noted previously, Skype uses what's known as a P2P architecture, wherein each computer—called a "node"—runs the same software and is treated equally. Some nodes in the network are designated "supernodes" and take on some of the tasks necessary to help organize and manage the network. Even though such nodes are designated "supernodes," they still run the same software as any regular node, and a node can become a supernode at any time, and vice versa. The important concept to understand is that all the computers in a P2P network run the same software and are treated equally.

When you start the Skype softphone running on your PC, it first connects to a Skype login server in order to authenticate you. After you're authenticated, the network broadcasts your online status—that is, makes your "presence" on the network known to others—and otherwise keeps track of your activities.

When you make a Skype call, Skype first locates the other party on its network, and then attempts to establish a direct link across the Internet between the two computers. By establishing a direct and dedicated link for communication, voice call quality is improved, as there are fewer things in the way to inhibit smooth and continuous data transfer. If a direct link cannot be established, the call is routed via other nodes in the network, but call quality can sometimes suffer as a result. Skype is both network and Internet savvy, so it will always do the best it can to ensure the most direct link possible in order to achieve good voice quality for a call.

Skype's P2P network contains millions of nodes and tens—or perhaps hundreds—of thousands of supernodes. In short, Skype's network is vast and spans the globe!

Some Advantages of Skype's P2P Network

Like all P2P networks, Skype continues to function well as it grows in size. As more and more users join Skype, they bring with them the processing power, data storage, and networking power needed to accommodate them. This is very different from traditional Internet applications, which require more and more servers and bigger and bigger network data pipes as more users join. In the traditional model of Internet applications adding more users eventually creates the need for more infrastructure, whereas, for a P2P network application such as Skype, additional users bring with them their own solution to growth!

Perhaps the most important consequence of Skype's growing, but self-sustaining, P2P network is that Skype doesn't have to invest in and maintain large data-processing facilities of its own. This means that Skype can provide its softphone and services at very low or no cost to its users. It also helps explain how Skype can afford to give away free phone calls without going out of business: the simple answer is that it costs Skype nothing to do so—you bring the PC, data storage, and network connection and Skype merely provides a piece of (free) software. That's why Skype can boast that Skype-to-Skype calls will always be free!

A Whirlwind Tour of Skype's Features

Skype comes with a lot of neat features for no more cost and effort than a free software download and simple install. These basic features—and there are many—can be further enhanced by subscribing to one or more of Skype's fee-based services.

All of the following features are thoroughly covered in this book.

Skype Free Features

Skype's free, basic features are as follows:

- **Skype-to-Skype calls:** Skype-to-Skype calls are free and always will be for noncommercial calls—regardless of location or duration! See Chapter 2 for details.

- **Video calls:** Skype allows you to see the other person during a voice call if both people have Skype-compatible video cameras and both have enabled this feature in Skype. With Skype video, you can talk face-to-face with someone else regardless of distance, and it won't cost you a penny! See Chapter 4 for details.

- **Toll-free calling:** Skype users can make free calls to most toll-free numbers, whereas some other Internet telephony providers charge for such calls. As an added bonus, you can even call toll-free numbers in other countries, which is something that regular phones often can't do!

- **Conference calls:** You and up to four other people can conference-call together at the same time. Moreover, during a conference call, participants can also chat together and transfer files to one another (or to all participants) by simply dragging-and-dropping files. See Chapter 2 for details.

- **Call forwarding:** Away from your PC, but don't want to miss any important calls? No problem, just use Skype's call-forwarding feature. Using call forwarding, Skype-to-Skype calls and SkypeIn calls from regular phones can be forwarded to as many as three Skype accounts or regular phone numbers (of course, you can only forward calls to regular phones if you are a SkypeOut subscriber, in which case the call rate appropriate to the call's destination will apply). See Chapter 7 for details.

> **Something Worth Knowing**
>
> If your PC has an Intel dual core processor in it, you can conference call with up to nine other people instead of the usual four. To determine whether your PC has a dual core processor, go to **Windows: Start: Control Panel** and in the window that appears double-click on **System**. Then, in the **System Properties** window, click on the General tab.

◆ **Chat:** Using Skype's chat feature you can exchange text messages with another Skype user in real time. See Chapter 2 for details.

◆ **Multi-chat:** Why limit yourself to one-on-one chat? With Skype's multi-chat feature you and a bunch of friends, family members, or work colleagues (up to 50 in total) can chat online in real time so that all text messages are seen by all chat participants. See Chapter 2 for details.

◆ **Chat emotional icons:** Otherwise known as "emoticons," these small icons can be inserted into text messages during chat sessions so that chat participants can quickly and easily express their emotional state or emotional response to chat activity in real time. See Chapter 10.

◆ **File transfer:** You can transfer a file to one or more Skype users from within your contacts list at any time or with anyone during an active Skype-to-Skype call or chat session. And you can do all this with little more than a right-click of your mouse, or by dragging-and-dropping a file.

Something to Try

You can select a collection of people from within your contacts list by holding down the control key on your keyboard while clicking on the contacts you wish to select with your mouse. As an alternative to selecting multiple entries in your contacts list, you can use the contact grouping feature of Skype, which is covered in Chapter 10.

◆ **Contact management:** Add your own contacts, search Skype contacts, import existing contacts held outside Skype, share your contacts with other Skype users, and group contacts into meaningful categories for convenience and to carry out group-specific tasks, such as conference calls, with a click of the mouse. See Chapter 10.

◆ **End-to-end encryption:** Skype uses encryption to scramble data so that it is unintelligible to any eavesdropper while it is being transmitted over the Internet. This means that Skype-to-Skype calls are more secure than regular or mobile phone calls! However, calls that use the public telephone network through SkypeOut and SkypeIn are not encrypted—and so are less secure than Skype-to-Skype calls—once outside the Skype network.

- **Privacy controls:** Control who can call you, who cannot call you, or who can chat with you. See Chapter 15.

- **Visibility controls:** Using Skype's online status settings you can control how visible, or invisible, you are on the Skype network. See Chapter 10.

- **Mood messages:** As part of how you are seen by others on the Skype network, you can choose to post a mood message that will be seen when another Skype user clicks on your entry in their contacts list. Mood messages can display any sort of short text message. See Chapter 11.

- **Skype toolbars:** Skype provides productivity-enhancing toolbars for Microsoft Outlook and Internet Explorer that more closely integrate those applications with Skype. See Chapter 12.

- **Skype buttons:** Using Skype buttons, you can add a clickable button to a web page, your e-mail signature, or your blog that allows others to easily interact with you though Skype. See Chapter 12.

- **Multi-platform support:** Skype runs on Windows, Mac OS X, Linux, and Pocket PC. This book focuses on Skype for PCs running Microsoft Windows.

- **Multi-language support:** The Skype softphone is available in 27 languages. See Chapter 11.

- **No spyware, adware, or malware:** More of a promise than a feature, this is a commitment by Skype to keep their software free from all other software that might spy on you, pester you with advertising, or otherwise maliciously interfere with your PC. See Chapter 14.

Skype Fee-Based Features

Briefly, Skype's fee-based features include the following:

- **SkypeOut:** Using SkypeOut, Skype users can call fixed and mobile phones at very low rates—in some cases, as low as 2¢ per minute. See Chapter 5.

♦ **SkypeIn:** SkypeIn provides a dial-in number that enables regular and mobile phone users to call Skype users. A SkypeIn number—and you can have as many as you want—costs $35 per year, or $12 for 3 months. See Chapter 6.

♦ **Voicemail:** Skype Voicemail ($6 for 3 months or $18 for a year) is for those times for when you are away from your PC or busy on another call and don't want to miss any incoming calls. As an added bonus, as a Skype Voicemail subscriber, you can send voice-mail messages to other Skype users, even if they aren't voicemail subscribers themselves! See Chapter 7.

♦ **Skype Zones:** If you want to use Skype while on the move, this might be for you. Skype Zones gives you access to more than 18,000 wireless Internet access points around the world from which you can make and receive Skype calls. In fact, you can use all the usual features of Skype, including any Skype services (SkypeOut, SkypeIn and Skype Voicemail) to which you have sub-scribed. See Chapter 8.

♦ **Skype Control Panel:** Skype Control Panel is targeted at busi-ness users of Skype who want to have a central administrator for several accounts, to simplify payment and management of Skype services. This service might also appeal to individuals who main-tain several Skype accounts. The Skype Control Panel and other business-oriented features of Skype are not covered in this book, but additional details can be found online at www.skype.biz.

♦ **Personalise Skype:** Using this service you can buy personalized pictures to add some pizzazz to how others see you online and buy custom ringtones that make Skype more pleasant and fun to use. See Chapter 11.

Alternatives to Skype

Skype isn't the only Internet-telephony game in town! Even though this book is primarily about Skype, you should also know that there are other Internet-telephony providers that are even cheaper than Skype for some types of call. For example, VoIP Buster offers free calls to regular landline phone numbers in the following countries: Andorra, Australia,

Austria, Belgium, Bulgaria, Canada, Chile, Colombia, Croatia, Cyprus, Denmark, Estonia, Finland, France, Georgia, Greece, Hong Kong, Iceland, Ireland, Italy, Japan, Latvia, Liechtenstein, Luxembourg, Malaysia, Monaco, Mongolia, Netherlands, New Zealand, Norway, Panama, Peru, Portugal, Puerto Rico, Singapore, Slovenia, South Korea, Spain, Sweden, Switzerland, Taiwan, Thailand, and Venezuela. In addition, you can even get a free regular dial-in phone number for select countries. Check the VoIP Buster website, www.voipbuster.com, for details and restrictions.

You might also want to check out some of the other alternatives to Skype shown in the following table. This list is by no means exhaustive, but it's a starting point for you to shop around for the best deal—and best features—in Internet telephony.

Some Alternatives to Skype

Softphone	Website
VOIP Buster	www.voipbuster.com
VOIP Stunt	www.voipstunt.com
Gizmo	www.gizmoproject.com
Free World Dialup	www.freeworlddialup.com
Woize	www.woize.com
Xten	www.xten.net
Google Talk	www.google.com/talk
Yahoo! Messenger	messenger.yahoo.com
Yahoo! Dialpad	www.dialpad.com
MSN Messenger	messenger.msn.com

The Least You Need to Know

◆ Skype is by far the most popular softphone for making free calls over the Internet.

◆ Skype's free services include voice calls, video calls, chat, and file transfers with other Skype users.

- You can enhance Skype by subscribing to fee-based services that, among other things, enable you to make calls to, and receive calls from, regular and mobile phones.

- Using Skype's cheap call rates, you can save a lot of money on long distance and international calls to regular and mobile phones.

- Alternatives to Skype exist, some of which can save you even more money for certain types of call.

Installing and Configuring Skype

In This Chapter

◆ Meeting Skype's PC system and Internet connection requirements

◆ Installing Skype

◆ Registering a Skype user account

◆ Configuring Skype

◆ Making your first Skype call

Skype has done a good job of making installation and configuration of its softphone program easy and trouble free. However, given the variety of PCs and peripherals in use today—not to mention the variety of other software running on people's PCs— these activities will never be entirely trouble free for everyone. In this chapter I help you install and configure Skype on your PC with minimal fuss, so you can start using Skype right away.

There are six steps to getting Skype installed and running smoothly on your PC:

1. Check that your computer hardware meets the minimum requirements to run Skype.

2. Check that your network setup is sufficient for, and compatible with, Skype.

3. Download and install Skype.

4. Set up a Skype user account (if you don't already have one).

5. Work through a checklist of configuration options for Skype that will help smooth the path to trouble free Skype calls.

6. Make a test call.

I discuss these steps in detail in the following sections.

PC System Requirements

Skype's computer system requirements are fairly minimal, as the following table shows. This means that, unless your PC is very old, you will most likely not have any problems running the Skype softphone.

Official System Requirements for Running Skype on a PC

PC Feature	Requirement
Operating system	Windows 2000 or XP
Processor speed	400 MHz
Memory	128 MB
Available disk space	15 MB
Audio input/output	Speakers and microphone
Video camera (optional)	USB webcam

If you want to run other programs while also using Skype, you will need additional processing power and memory beyond what's indicated in the previous table. Likewise, if you want to use Skype's more advanced features, such as conference calling or video, you will need additional processing power and memory. Each new release of the

Skype softphone introduces new features and enhancements, and these normally require more processing power and memory. In other words, the more processing power and memory you have, the better your overall Skype experience will be.

Something Worth Knowing
Even though Skype doesn't recommend its softphone for use on Windows 98 and Windows ME, many PC users run Skype on these platforms without problem. Skype won't work as well as it does on Windows 2000 or Windows XP, and some features won't work as advertised; but it might be worth a try if Windows 98 or Windows ME is your only option. Note that Skype video is supported only on Windows XP.

This is not to say that Skype won't run well on a machine that only meets the minimum requirements. It just means it won't run as well as it would if you had processing and memory resources to spare. If you want a smooth and pleasant experience when using any of Skype's current features, I suggest you consider the requirements in the following table as your target.

Author's Suggested System Requirements for Running Skype on a PC

PC Feature	Requirement
Operating system	Windows XP
Processor speed	1.2 GHz
Memory	256 MB
Available disk space	30 MB
Audio input/output	USB headset or handset
Video camera (optional)	USB webcam

To find out the specification of your PC, go to **Start: Control Panel,** double-click on **System** to open the **System Properties** window, and select the **General** tab. A sample System Properties window appears in the following figure. Compare the information in the System Properties General tab to the requirements specified in the previous tables.

The System Properties window.

Operating system

Processor

Memory

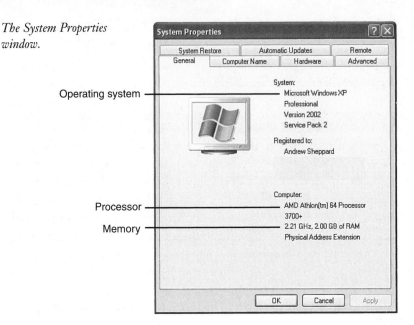

Provided your PC meets or exceeds Skype's minimum requirements, you are ready to move on to see whether you meet Skype's network requirements.

Internet Connection Requirements

Skype can work with a fast dial-up Internet connection over a telephone line in a pinch, but don't expect much in the way of quality. For practical day-to-day use, you have to have a *broadband* connection to the Internet, such as that provided through *cable* or *DSL*.

def•i•ni•tion

Broadband is normally taken to mean an Internet connection having a data rate greater than 128 kilobits per second (Kbps), and is often substantially higher, perhaps a few megabits per second (Mbps). For comparison, the maximum data rate for a dial-up connection is 56 Kbps.

Cable and **DSL** (the latter standing for Digital Subscriber Line, which also has a close cousin called Asymmetric DSL, or simply ADSL) are the two most common technologies used to provide broadband connectivity to the Internet for the home.

To get the most out of Skype, there are two characteristics of your broadband Internet connection that you must pay attention to:

- ◆ **Bandwidth,** measured in bits per second, is a measure of the rate at which data can be transmitted over the connection. The higher the bandwidth, the better your Skype experience will be, as your voice will be clearer during a call.

- ◆ **Latency,** measured in milliseconds, is the delay between when you start speaking during a call and when the person at the other end hears your words. If this delay is too long, say, more than half a second, the other person might start talking back to you before you are done talking. A long latency means that conversations will be stuttering and awkward, as each person runs the risk of talking over the other. The shorter your latency, the better will be your Skype experience.

Most cable and DSL Internet connections are asymmetric in the sense that the rate at which you can send data is different from the rate at which you can receive data. That is, the bandwidth in each direction is asymmetric. Normally the rate at which you can send data is substantially less than the rate at which you can receive data. From Skype's point of view, the overall quality of a voice call over the Internet will be limited by the minimum of the two bandwidths in either direction: send or receive.

To find out the bandwidth and latency for your Internet connection visit www.numion.com and click on the **YourSpeed** link, then click on the **Quickstart** link. This will run a test that will tell you your bandwidth for both send and receive, and tell you the latency for your Internet connection. For a good experience with Skype, you will need a bandwidth of at least 128 Kbps for both send and receive, and a latency less than 500 milliseconds (that is, a latency less than half a second).

Downloading and Installing Skype

To download Skype, visit www.skype.com in Microsoft Internet Explorer (IE) and click on the **Download** menu item at the top of the home page. This displays the Download Skype page. Click on the link for the **Windows platform,** which takes you to the download page specifically for Skype for Windows. Next, click on the **Get it now** button, which displays a page with helpful step-by-step instructions. Click on the **Download** button, which will then display the popup window shown in the following figure. This popup gives you three ways to proceed: **Run, Save,** or **Cancel.** The third option, Cancel, is fairly obvious in that it cancels the download process. Run and Save can both be used to run the Skype setup program, but they do so in different ways, which I explain in the following sections.

Something Worth Knowing

Before downloading and installing the Skype softphone, you might want to familiarize yourself with the user interface—the look and feel of the Skype softphone. That way, when you run Skype, you will already know how to do some basic things with it. Check out the screenshots of the Skype softphone at www.skype.com/download/screenshots.html. Also, look at the online user guides for Skype at www.skype.com/help/guides.

File download window for the Skype setup program.

Downloading and Running the Skype Install Program from Within Internet Explorer

Clicking on the **Run** button will download the Skype install program to your PC. After it is downloaded, you'll see the window shown in the next figure. Click the **Run** button in this popup to start the Skype install program. If you don't want to install Skype, click **Don't Run** to cancel the install process.

Run download window.

Downloading the Skype Install Program from Within Internet Explorer and Then Running It

Clicking on the **Save** button will generate the popup window shown in the next figure. Using this window you can choose a folder in which to save the Skype install program. Be sure to remember where you put it, as you'll need this piece of information for the next step.

Save file window.

After the Skype install program has finished downloading, you must run it. The easiest way to do this is to navigate to the Skype install program using the Windows Explorer file browser, and then double-click on it. This will run the Skype install program, though you may be asked by Windows to confirm that you want to run this program.

Running the Skype Install Program

When you run the Skype install program, you will first be presented with an introductory window that enables you to choose your preferred language, as shown in the figure that follows. This will select the language used for both the remainder of the installation process and the default language for operation of the Skype softphone (to change the preferred language after installation, see Chapter 11).

The opening window of the Skype install program.

Clicking the **Next** button in the Setup window takes you to a window that asks you to accept (or reject) the Skype license agreement, as shown in the next figure. Select "I accept the agreement" and click **Next** to proceed to the final step of the installation process. (Selecting "I do not accept the agreement" aborts the installation.)

Accept (or reject) the Skype license agreement.

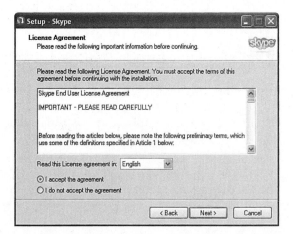

Click the **Finish** button in the final window of the Skype setup to launch the Skype program.

Finish the installation of Skype.

Something Worth Knowing

Skype issues frequent updates of its softphone. The latest version of the Skype softphone that is available for download is shown on the download page at www.skype.com. It is to your advantage to keep your Skype softphone current.

To find out what version of Skype you are running, go to **Skype: Help: About** in the Skype program. This displays a splash screen with the version number of the Skype softphone you are running.

Register a Skype User Account

If you don't already have a Skype user name and account, you'll have to set one up. This will give you access to all of Skype's free services (remember, it's completely up to you whether you want to subscribe to any of Skype's fee-based services).

Start Skype by double-clicking on its icon on your desktop, or from the Windows menu (go to **Start: All Programs: Skype: Skype**). This presents you with the **Sign In** window shown in the next figure.

Skype sign in window.

Click on the link **Don't have a Skype Name?**, which displays the window shown in the next figure.

Create Skype account window.

Fill in the details for the **Create a new Skype Account** window, and then click on the **Sign In** button. You are now signed in to the Skype network, and you should see the Skype softphone displayed on your screen.

Something Worth Knowing

Your Skype user name is the primary means by which others on the Skype network will get to know you, and an inopportunely chosen name might attract a lot of unwanted attention. Even though Skype names, once created, cannot be changed or deleted ever, you can register any number of Skype names, provided they're available. So a badly chosen name isn't any great disaster. However, before choosing a new Skype name you might want to briefly skip ahead to Chapter 15 for some advice on the topic.

You'll use your Skype name and password to sign in to the Skype website to manage your account and sign up for Skype's fee-based subscription services.

Basic Configuration of Skype

Before you start using Skype, I encourage you to review the following checklist. This list covers problem areas and gotchas that often trip up new users of Skype. By casting your eyes over a short list of Skype configuration settings, you will most likely avoid any unpleasant surprises. So start Skype running, sign in, and check these items before making your first Skype call.

❏ **Skype: Tools: Options ... : Privacy**   Check your privacy settings. Specifically, you can choose from whom you would, and would not, like to receive calls and chat.

❏ **Skype: Tools: Options ... : Sound Devices**   Set your Audio In and Audio Out sound devices to the device or devices you wish to use with Skype. I do not recommend using the setting "Windows default device."

❏ **Skype: Tools: Options ... : Connection**   Make sure there's a check mark opposite "Use port 80 and 443 as alternatives for incoming connections."

❏ **Skype: Tools: Options ... : Advanced**   Put a check mark opposite "Start Skype when I start Windows," so that Skype will always be available when you're using your computer.

Having worked through the checklist, you are now ready to make your first Skype call. But no matter who you call, you will no doubt want to impress them, right? Well, if so, perhaps there's one last thing to do: make a test call.

Making Your First Skype Call

Skype has conveniently provided a call test service that enables you to test both your Skype softphone and your sound devices, ensuring that during a call others will be able to hear you, and you will be able to hear them.

To make a test call, click on the **Contacts** tab in the Skype softphone. Select the contact named **Skype Test Call,** and then click on the large round green button with a phone on it, as shown in the next figure. This connects you to the Skype test call service.

Making a call to the Skype test call service.

During a Skype test call, an automated voice at the other end of the call gives you instructions on how the test works. But in essence, the test consists of you talking to the test service and the test service playing back your voice to you. If you hear your own voice clearly, everything is working and you can start using Skype. Congratulations! You've just made your first Skype call.

> **Something Worth Knowing**
>
> If you can't get the Skype test call service to work, you might want to jump ahead to Part 5 of this book, which shows you how to troubleshoot Skype.

Making a Conference Call, Chat, or File Transfer

Before leaving this chapter, the more adventurous reader might want to quickly test some of the other features of Skype: conference calling, chat, and file transfer. To do that, you'll need to have a handful of people setup in your contacts list. So get the Skype names of some friends (or use the Skype search feature [**Skype: Contacts: Search for Skype Users** ...]) to populate your contacts list, and then try the following quick experiments. They'll take only a few minutes and will be time well spent.

To make a conference call with Skype, select two or more people in your contacts list by holding down the Control key on your keyboard while clicking on the names with your mouse. Next, click on the **Conference Call** button on the toolbar (see Appendix D if you need some help locating the button), and the conference call should start in its own call tab.

To start a chat session with someone, click on their name in your contacts list and then click on the **Chat** button on the toolbar (again, use Appendix D if you need help identifying the correct button). This opens Skype's chat window and starts a chat session. To start a multi-chat session, just highlight more than one name in your contacts list (by holding down the Control key on your keyboard while clicking on their names with your mouse) and click on the **Chat** button.

> **Something Worth Knowing**
>
> Conference calls, multi-person chat, and multi-person file transfers are made easier if you can work with contact groups, or lists of people, rather than individual names in your contacts list. This is a topic covered in Chapter 10.

Finally, try sending a file to someone. Click on a name in your contacts list and then click the **Send File** button on the toolbar. This opens a window which you can use to navigate to, and select, the file you want to send. Pick a file —any file—and click on the **Open** button. This initiates a file transfer. You can send a file to more than one person at the same time by selecting multiple names in your contact list before clicking on the **Send File** button.

The Least You Need to Know

◆ Your PC must meet some minimum requirements to run the Skype softphone.

◆ Your Internet connection must meet the needs of the Skype softphone in order to work.

◆ Skype has made the installation process simple and trouble free for most PC users.

◆ You can set up a Skype name and user account for free—all you need to do so is a user name and password.

◆ By checking a few things immediately after installing Skype and before using it, you can avoid some common problems.

◆ Skype makes testing your setup a snap by providing a test call service.

Chapter 3

Navigating Skype's Graphical Interface

In This Chapter

- ◆ Getting to know Skype's graphical interface elements
- ◆ Looking more closely at Skype's most frequently used elements
- ◆ Navigating Skype's interface
- ◆ Customizing the graphical interface

Skype's softphone has gotten a lot of praise for the simplicity and clean lines of its graphical user interface, particularly considering how many things Skype can do for you. Even so, if you are just starting out with Skype, you'll have to learn a whole new way of doing things and a whole new set of terms. The aim of this chapter is to help you get up to speed as quickly and as painlessly as possible.

Skype's Graphical Elements

Skype's main window, as the next figure shows, contains quite a few graphical elements.

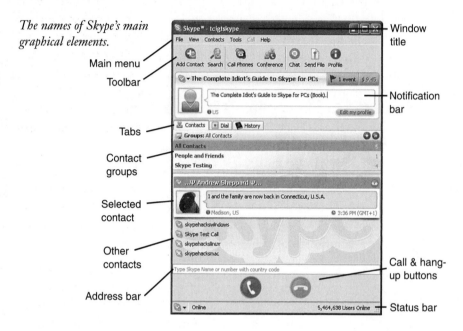

The names of Skype's main graphical elements.

In brief, here's what each of Skype's major graphical elements does:

- ◆ **Window title:** Displayed in the title bar of Skype's main window is the Skype username for the person who is currently signed in to Skype. In the accompanying figure, the name is "tcigtskype", which is short for—no prizes for guessing correctly—*The Complete Idiot's Guide to Skype!* This is the Skype username that will be used throughout this book. Yours will be different.

- ◆ **Main menu:** The main menu is what gives you the most control when using or configuring Skype.

- ◆ **Toolbar:** From the toolbar you can do things quickly and efficiently by clicking on an icon.

- ◆ **Notification bar:** This bar is divided horizontally into four parts, from left to right: Online Status (as a small pulldown menu),

Full Name (if you've set one up, otherwise your Skype name is displayed), Events, and My Account. Clicking on the status portion of the bar drops down a menu from which you can set your online status. Clicking on your name drops down a text box, into which you can enter a short message that others will see when you are online. Clicking on Events drops down a box that shows all events, such as missed calls, that happened while you were away from Skype. The portion of the notification bar to the far right is labeled "My Account" if you are not yet a SkypeOut subscriber; if you are already a SkypeOut subscriber, it displays the value of your remaining SkypeOut credit balance. Clicking on this portion of the notification bar drops down a box that shows your subscription status to Skype services.

◆ **Tabs:** The three tabs shown can be used to display the Contacts, Dial, and History panels, respectively. These three panels are dealt with in more detail in the pages that follow. When you make or receive a call, a new panel (and tab) will appear; again, this will be described in more detail in the following pages.

◆ **Contact groups:** For your convenience, you can group your contacts into either predefined categories or categories you define yourself. Using this feature, you can quickly filter your contacts list to isolate only those contacts of interest. Moreover, by using contact groups, rather than individual entries in your contacts list, you can communicate more easily when you need to interact with many people at once. Note that contact grouping has to be enabled from the main menu: go to **Skype: View** and check **Show Contact Groups.**

◆ **Selected contact:** When you select a contact in your contacts list by clicking on it, it expands to show additional information.

◆ **Other contacts:** Contacts that are not selected show only their Skype username or a name you have assigned to them for display purposes in your contacts list. You can also show additional information about any contact by letting your mouse pointer hover over that contact.

Something Worth Knowing

Even if you are not a SkypeOut subscriber, you can still call many toll-free numbers, including many toll-free numbers in other countries. (Be sure to prefix them with the correct country code.)

- ◆ **Address bar:** You can enter a Skype username or toll-free number or a regular telephone number (if you are a SkypeOut subscriber) in the address bar to start a call.

- ◆ **Call & hang-up buttons:** Use the green call button to start a call; use the red hang-up button to end a call.

- ◆ **Status bar:** On the far left of the status bar is a pull-down that enables you to set your online status, which is covered in detail in Chapter 10. The number on the far right of the status bar indicates how many users are signed onto Skype. This number has grown, and continues to grow, at a phenomenal rate.

This list is meant to give you an overview of Skype's graphical user interface. Appendix D lists the names and the purpose of all Skype buttons and icons. For now, let's take a closer look at some of the elements you'll be using frequently.

Tabbed Panels

When there is no call in progress, Skype usually has three tabbed panels—Contacts, Dial, and History—in its main window. Only one is displayed at any given time, and you can switch between panels by clicking on the tabs.

Clicking on the Contacts tab displays your contacts list panel, as shown in the following figure. In this panel you can select an existing contact to call, chat with, or send a file to. If you have contact grouping feature turned on, you can quickly filter your list of contacts by category. As a further aid to finding a contact, in this mode the address bar acts as a filtering tool, so that as you type a Skype username (or the name you have assigned to a user for display purposes) the closest match in your contact list is highlighted.

Clicking on the Dial tab displays Skype's dialpad, which is shown in the next figure. The dialpad emulates the functions of a keypad that you might find on a regular or mobile phone. Clicking on the buttons of Skype's dialpad enters the digits and symbols (*, +, and #) into the address bar. You can also use the keyboard to enter names and numbers into the address bar.

Contacts list panel.

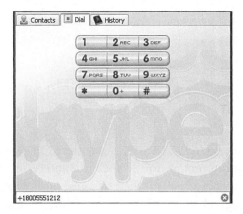

Skype's dialpad panel.

Clicking on the **History** tab displays a history of Skype events (missed calls, incoming calls, outgoing calls, voicemails, transferred files, and chats), as shown in the following figure. When the history panel is displayed, a small pull-down menu (next to a small picture of a magnifying glass) appears on the far left of the address bar. Use this pull-down menu to filter the event history to show only those events of interest.

Something to Try

Provided "quickfiltering" is enabled (go to **Skype: Tools: Options: Advanced** and put a check mark against "Enable Contact List and History quickfiltering"), typing a Skype username or number in the address bar when either of the Contacts or History tabs are active will filter the list and highlight the closest matches as you type.

Skype's event history panel.

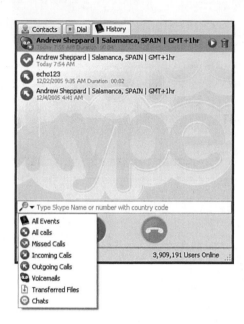

Call Panel

When you make or receive a call, Skype adds a new tab panel, as shown in the following figure. In the center of the panel is a picture of the other party of the call, if they have set up their account to show their own picture; otherwise a default picture provided by Skype is shown. In the case of a conference call, pictures for all other parties to the call are tiled in the call panel. The duration of the call is displayed at the bottom of the call panel.

Skype's call panel.

Chat Window

When you start a new chat session or participate in an existing session, Skype opens a window, as shown in the next figure. Using chat, all participants can contribute text messages that will be seen by all other chat participants.

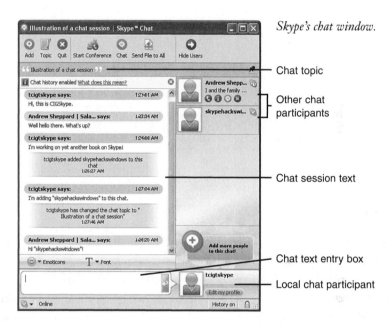

Skype's chat window.

Chat topic

Other chat participants

Chat session text

Chat text entry box

Local chat participant

Video Panel

When you make or receive a video call, Skype adds a new tab panel, as shown in the following figure. Participants who have a *webcam* will be shown in the video panel; any video-call participants who don't have a webcam will have their pictures (if they have one) displayed instead.

Even though a video call will begin in a tab panel like that shown in the following figure, you have the

def•i•ni•tion

A **webcam** is a video camera that attaches to your PC, usually through a USB port, and provides real-time video input for applications that run on your PC. For Skype video calls, the webcam should be mounted in such a way that it shows your face, so the person at the other end of the video call can see you while you speak.

option of displaying video in a separate and larger window, or making the video you see full screen.

Skype's video panel.

Navigating Skype's Graphical Interface

The best way to learn to navigate Skype's graphical user interface is by doing things. In this spirit of learning by doing, what follows are several common tasks that you might want to do in a how-to format.

How to Make a Call to a Skype User

If the person you want to call is already in your contacts list, the simplest way to call him or her is to go to your contacts list (click on the Contacts tab to display the contacts list panel), select the contact you want by clicking on it with your mouse, and then click on the call button. This starts a call and opens a tabbed call panel for that call. Alternatively, once you have selected the contact, rather than clicking the call button, you can either double-click on the selected contact, or right-click on the selected contact and pick "Start Call" from the popup menu that appears.

If you want to call someone who is not in your contacts list, you must first enter his or her Skype username in the address bar, and then hit the enter key on your keyboard or click on the call button. This will open a tabbed call panel for that call.

How to Search for Other Skype Users

You can search for other Skype users in two different ways. First, you can click on the **Add Contact** button on the toolbar, which opens a new window in which you can search for a Skype user by entering that person's real name, Skype name, or e-mail address. Second, by clicking on the **Search** button on the toolbar, a window opens that allows you to search for a Skype user by entering that person's real name, Skype name, or e-mail address; but in addition, you can search using these criteria: country, state, city, language, gender, and age range; and you can also search for people whose online status is in Skype Me mode.

> **Something Worth Knowing**
>
> The default behavior for double-clicking on a person in your contacts list or event history list is to start a voice call with that person. However, this double-click behavior is configurable; go to **Skype: Tools: Options: General** and toggle the radio buttons between "Start a call" and "Start a chat" for "When I double-click on Contact or use the address field").

How to Start a Multi-Person Chat Session

Click on the **Contacts** tab to display your contacts list. While holding down the control key on your keyboard, click on the Skype users with whom you want to conduct a chat session. With multiple Skype users thus selected, click on the **Chat** button of the **Skype toolbar;** or right-click with your mouse and choose **Start Chat** from the popup menu that appears. Starting a chat session in this manner will open a chat window with all the selected Skype users as active participants.

Customizing Skype's Graphical Interface

You can choose to hide many of the graphical elements described in this chapter. Taken to its extreme, you can make Skype's graphical user interface very minimal indeed, as the following figure shows. The important thing to keep in mind is that Skype gives you control over how it looks and behaves. To change the look and feel of your Skype softphone, experiment with the settings located under the main menu (go to **Skype: View**).

A minimal user interface for Skype.

Finally you can change the language used to annotate Skype's graphical elements. Go to **Skype: Tools: Change Language,** and select your preferred language.

The Least You Need to Know

◆ Skype has a compact, but simple graphical user interface.

◆ By learning the names and purpose of Skype's graphical elements you will be able to more easily navigate and use Skype's softphone.

◆ The best way to learn how to use Skype is to experiment with its features.

◆ You can customize the Skype softphone by selectively hiding graphical user interface elements.

Part 2

Advanced Skype and Skype Services

Now that you've got the hang of Skype-to-Skype calling, it's time to tackle Skype's more advanced features: using Skype Video to make video calls and SkypeOut to call regular and mobile phones. In addition, I'll show you how you can use SkypeIn to set up one or more regular dial-in numbers so non-Skype users can call you, and how to use Skype's voicemail and call-forwarding features so that you never miss a call. Finally, I show you how, by using these advanced features and services of Skype, you can shave dollars off your phone bills!

Chapter 4

Skype Video

In This Chapter

- ◆ Learning about Skype video
- ◆ Meeting minimum requirements for Skype video
- ◆ Setting up a webcam
- ◆ Configuring Skype video
- ◆ Testing Skype video

Skype video enables you to exchange live video images in addition to voice. That means that you and the person you are calling can see each other while you talk. And, best of all, Skype video calls—like their voice counterparts—are free!

To take advantage of Skype video, both you and the person you are calling must be Skype users, and each of you must be set up with a Skype-compatible webcam. If only one of you has a webcam, or if only one has Skype video enabled, then the video will only work in one direction. Your computer must also be able to cope with the demands of video, as must your Internet connection.

Skype Video Requirements

For you to make use of Skype video, you need to meet certain minimum requirements, as detailed in the following sections.

PC Requirements

In Chapter 2, I covered the general requirements for your computer to run Skype. In that chapter I encouraged you to treat Skype's minimum system requirements as exactly that, minimum requirements. If you want to have a good experience using video, you should double those minimum requirements. Skype recommends, as a minimum, that your processor speed be around the 1 GHz mark. As most webcams connect to your PC using *USB*, you will also need a spare USB port.

def•i•ni•tion

USB is short for universal serial bus. It's a common method by which to attach various peripherals to your PC, such as a headset, handset, or webcam.

Internet Connection

I outlined the minimum broadband Internet requirements for voice calls in Chapter 2. Skype video, if you are to have anything like real-time moving pictures, puts additional demands on your Internet connection. Specifically, you will need a good deal more available bandwidth (data-transfer speed) than for plain-vanilla Skype voice calls. Skype recommends a bandwidth (send and receive) of 256 Kbps for Skype video to work well. Skype video will work at lower speeds, but the streaming image might be jerky and broken up.

Something Worth Knowing

You can find a list of Skype-compatible webcams at forum.skype.com/viewtopic. php?t=41162. If you don't already have a webcam and are planning to buy one, I encourage you to check this list first, or search the forums using the name of a specific webcam.

Webcam

To use Skype video, you will need a Skype-compatible webcam. You will also need to install the webcam and configure it for use by Skype. Note that some webcams come equipped with a built-in microphone, which might be useful if you don't yet have a microphone or USB headset, or if you prefer to talk to the webcam during a call (that way, you're more likely to look into the webcam). However, in my experience, it's a far more enjoyable experience to use the webcam as only a camera and to use a good USB headset (or handset) for audio.

Setting Up a Webcam

A typical webcam plugs into a USB port on your PC and comes with drivers and software to make it work with Windows. Some webcams require the drivers and software to be installed on your PC before you plug in your webcam, while others do it the other way around. If you follow the instructions supplied by your webcam manufacturer, you should be fine.

To install a webcam, you will need a spare USB port. Most PCs have anywhere from two or four USB ports, but given the number of peripherals that use USB, these can fill up pretty quickly. If you need more USB ports for your PC, you can use a *USB hub* to give you additional ports.

To use the webcam to make Skype video calls, you will need to position it so that you can look at it while you talk. Most webcams come with a privacy shield, which can be placed over the lens to block your image when you don't want anyone to see you. Many webcam users leave the privacy shield in place most of the time, and only flick it out of the way when they're sure they want to be seen.

def•i•ni•tion

A **USB hub** is a device you can use to expand the number of available USB ports for your computer. You plug the USB hub into a USB port of your computer. Typically there are between 4 and 8 USB sockets on the USB hub.

Configuring Skype Video

Go to **Skype: Tools: Options ...**, and click on **Video** to display the configuration settings for video, as shown in the following figure. Most of the configuration settings in this figure are self explanatory and, as you can see, Skype video is enabled by default when you install Skype. If you don't want to make video calls (perhaps your bandwidth isn't quite up to it, or you just don't like it), you can uncheck **Enable Skype Video.** Alternatively, you can leave Skype video enabled, but uncheck **Start my Video Automatically,** which gives you control over when to make your video available to be seen by others.

Configuration settings for Skype video.

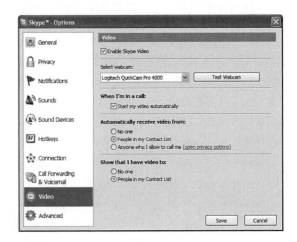

Testing Skype Video

When testing Skype video, you should first test your webcam to make sure it works. You can then call someone who has a webcam and who is willing to speak with you.

To test your webcam, click on the **Test Webcam** button in Skype video settings. This should open a window that, if all is working properly, shows your video image, as in the following figure. Clicking on the button labeled **Webcam Settings** displays a window that enables you to make adjustments (brightness, contrast, and so on) to the settings of your webcam. Change these settings to optimize your video image.

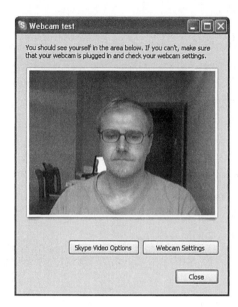

Skype webcam test.

To run a full-blown test of Skype video, you'll need to find someone—either in your contacts list, or in the general Skype community—who has a webcam set up for Skype and is willing to have a video call with you. If you're completely new to Skype and don't know anyone who uses Skype Video, you can try to find someone in Skype-Me! mode who is willing to speak and video with you. People in Skype-Me! mode typically welcome unknown callers, and such people can be found using Skype search (click on the Search button on the toolbar), or by visiting the Skype forums at forum.skype.com/index.php and scrolling down to the Skype-Me! listing.

The Least You Need to Know

- Skype supports Skype-to-Skype video and voice calls. Both are free!

- Before attempting to make Skype video calls, make sure your PC and Internet connection are up to the job.

- If you don't already have one, you'll need a Skype-compatible webcam to make video calls.

- After you configure and test Skype video with your webcam, you can make and receive video calls from other Skype users.

Chapter 5

Using SkypeOut to Call Regular Phones

In This Chapter

◆ Learning about SkypeOut

◆ Getting SkypeOut

◆ Buying more SkypeOut credits

◆ Rounding your calls

◆ Avoiding higher SkypeOut rates

SkypeOut is a fee-based service that connects the Skype P2P telephone network (which runs on the Internet) to the regular phone network. Using SkypeOut, you can call regular landline and mobile phones, mostly at very low per-minute call rates.

To use SkypeOut you must buy credits from Skype. SkypeOut credits are prepayments against future calls to regular and mobile phones. Each time you make a SkypeOut call, the call duration (measured in minutes) times the per-minute call rate for the call destination will be debited from your credit balance. In this sense, Skype credits are similar to prepaid phone-calling cards.

Getting SkypeOut

The easiest way to become a SkypeOut subscriber is to click on the **My Account** tab of the **Notification Bar** in the Skype softphone. This displays a group of links, including one for SkypeOut, as shown in the following figure. Click on the link for **SkypeOut.** This opens a browser window in which you are asked to sign in to your Skype account using the same user name and password that you use to sign in to the Skype softphone. After you've signed in you will be presented with a web page giving an overview of your account. Click on the link **Buy Skype Credit.** As a new Skype user your only option, using the **Buy this** button, is to buy 10 ($12) of credit to activate SkypeOut and provide some initial call time.

> ## Something Worth Knowing
>
> You don't have to be a SkypeOut subscriber to call certain toll-free numbers. For example, any Skype user can call 1-800-GOFEDEX (or equivalently 1-800-463-3339), without being a SkypeOut subscriber and without paying for the call. As an added bonus, you can even call many toll-free numbers outside the United States for free—try that with a regular phone!

Signing up for SkypeOut.

After clicking on the **Buy this** button for 10 € ($12) of SkypeOut credit, you are asked to verify some information, including your e-mail address and Skype password. You must also select a preferred currency for payment purposes. Finally, you must agree to Skype's terms of service and click the **Submit** button.

Skype will process your order and send a confirmation code—normally within just a few minutes—to the e-mail address you provided. Enter this confirmation code into the current web page, and then click the **Submit Confirmation Code** button. Make sure your preferred currency is shown, and then click the **Buy this** button. This takes you to a web page form that asks for your name and billing address. Fill in the form with your details. There are two buttons at the bottom of this form. The first is labeled **Save and Use It Now,** and the second is labeled **Use This Address Only Once.** If you plan to buy more SkypeOut credits in the future, click **Save and Use This Now.**

Something Worth Knowing

Skype is a European company, and its services are priced in the currency of the European Union, the euro, symbolized by €. The exchange rate between the U.S. dollar and the euro varies. At the time this book went to press, 1€ equaled US$1.20, and this is the rate I have used throughout this book when converting from euros to dollars. However, bear in mind that by the time you read this book the exchange rate might be different than what's quoted here.

By this time you must be wondering whether there's an end to this process! The registration process has been set up to combat fraud and, fortunately, you only have to go through it once. So, hang in there—you're almost done!

At this point you should be at the payment method page. Pick your preferred method of payment and click the **Continue** button. This takes you to the appropriate web page, operated by one of several financial institutions that process payments for Skype, based on your chosen method of payment.

Follow the instructions for your chosen payment method. After your payment is processed, you should see that SkypeOut has been activated and your account balance shown in the following figure. You are ready to start using Skype to call regular and mobile phones!

Something Worth Knowing

Whereas you can register your Skype name and use Skype's free services without providing an e-mail address, in order to use SkypeOut (or any other of Skype's fee-based services) you will need to provide Skype with a valid e-mail address.

SkypeOut activated and your credit balance.

SkypeOut Antifraud Restrictions

You should be aware of some of the antifraud restrictions Skype has put in place for new SkypeOut subscribers. The first three months as a SkypeOut subscriber are a probationary period in which the following restrictions apply:

◆ You will only be able to add to your SkypeOut credit balance if it falls below 5 € ($6), or 50% of your initial purchase.

◆ If you pay by credit card, you will be limited to buying a total of 20 € ($24) SkypeOut credits per month.

◆ If you pay using PayPal, your limit will be 30 € ($36) per month.

◆ Using other payment methods, you are limited to 100 € ($120) per month.

After three months, the restrictions are lifted. However, there is another restriction that never goes away: once you use a payment method (for example, a credit card or PayPal account) for one Skype account name, it cannot be used for any other Skype name. If you try to use that same credit card or online account to buy SkypeOut credits or other services for another Skype username, payment will be rejected. You can have several alternative payment methods for a single Skype account; it's just that those same payment methods cannot thereafter be used for any other Skype account.

Buying More SkypeOut Credits

Adding to your SkypeOut credit balance is a lot easier than the initial activation process. Just click on the **Your Skype Account is Activated** link, as shown in the previous figure. This opens a browser window and displays the sign-in page for your account. Sign in using your Skype username and password. This takes you to the Overview page for your account. Click the **Buy More Skype Credit** link and follow the payment instructions.

Caution

Your SkypeOut credit balance will expire and become useless after 180 days (six months) of inactivity. To keep it alive and kicking, all you need to do is make one SkypeOut call every 180 days.

Call Rounding

SkypeOut calls are rounded up to the next whole minute. So, if you make lots of short calls, you will end up paying for a lot of fractions of unused minutes. This "call rounding" effect means that your effective per minute call rate can be somewhat higher—up to about 40 percent more—than the advertised per minute call rate.

However, the call rounding effect falls off quickly as typical call durations increase, as the following table shows:

Unused Call Time as a Result of Rounding up to the Whole Minute

Typical Call Duration	Average Unused Call Time (approx.)
15 seconds–1 minute	38 percent
3 minutes	22 percent
10 minutes	9 percent
15 minutes	6 percent
30 minutes	3 percent

Even with the call rounding effect, SkypeOut's low per-minute call rates ensure that most of your calls will nevertheless be cheap. At Skype's lowest global calling rate of $0.021 per minute, the worst case effect of call rounding results in an effective per minute call rate of only $0.029, which is still cheap when compared with most rates for domestic inter-State and international calls of regular and mobile phone plans.

Something Worth Knowing

SkypeOut calls of four seconds or less are free, and the clock starts running once a call connects. If you bail out of a SkypeOut call within the first four seconds, it will cost you nothing. So, rather than wait for someone's answering machine, hang up and save yourself some money! Note that only the first four seconds of the very first minute of call time are free. After four seconds, all calls are rounded upwards in increments of one minute.

Avoiding Higher SkypeOut Rates

Many SkypeOut calls can be made at Skype's low global rate of $0.021 (0.017 €) per minute. However, SkypeOut rates do vary by destination and the type of phone being called, and they are often more expensive when a call is made to a mobile phone. Before calling a new number, you can check the SkypeOut rate by using Skype's Dialing Wizard, at www.skype.com/products/skypeout/rates/dialing.html. Using the Dialing Wizard you can check the SkypeOut rate for a number before making the call, which can avoid a costly mistake, as a handful of SkypeOut rates are $1.20 (1 €) or more per minute!

The Least You Need to Know

- ◆ SkypeOut allows you to call regular landline and mobile phones from Skype.

- ◆ SkypeOut is a fee-based subscription service that you must sign up for.

- ◆ You can add credits to your SkypeOut balance when your balance becomes low or is exhausted.

- ◆ For new Skype users, antifraud restrictions limit how many, and how often, SkypeOut credits can be bought. These restrictions disappear after three months.

- ◆ The call-rounding effect makes short duration calls more expensive than the advertised rate.

- ◆ You can check a SkypeOut call rate to a particular number before making a call by using the Skype Dialing Wizard.

6

Using SkypeIn to Receive Calls from Regular Phones

In This Chapter

◆ Learning about SkypeIn

◆ Picking a SkypeIn number

◆ Signing up for SkypeIn

SkypeIn is a fee-based subscription service that enables people using regular and mobile phones to call Skype users. In effect, it gives your Skype softphone a regular dial-in number; or, indeed, dial-in numbers, as a single Skype account can have up to 10 numbers. By dialing your SkypeIn number from a regular or mobile phone, anyone can connect to your Skype softphone.

Perhaps the biggest convenience factor of having a SkypeIn number is that it goes where you go, as calls are routed to wherever you happen to be running Skype and are signed in. For example, if you're in Hong Kong on business and have an

Internet connection, then calls made to your U.S. SkypeIn number will ring in Hong Kong. Whoever is calling your U.S. number will pay only the appropriate call rate for that number's area code, while the call will be routed to Hong Kong over Skype's network for free. Total cost for you as the recipient of the call is zero, as you pay only a fixed subscription fee to have the SkypeIn dial-in number, regardless of the number of calls you receive.

Picking a SkypeIn Number

When you get a regular phone number from your local telephone company, you are typically only given the option of getting a number with the local area prefix of your geographic location. That is, you are restricted to dial-in numbers for the area where you live.

This is not the case with SkypeIn numbers. With SkypeIn there's nothing stopping you from having a London, England, number, even if you live in New York. Even better, there's nothing stopping you from getting SkypeIn numbers for London *and* New York; and, based on your needs, dial-in numbers dotted around the globe. Welcome to the new world of Internet telephony!

At the time of this writing, SkypeIn numbers were available for the following countries: United States, United Kingdom, Denmark, Estonia, Finland, France, Germany, Poland, Sweden, Switzerland, Brazil, and Hong Kong. Within each of these countries you are typically given the choice of several area codes to choose from.

You can choose any country and available area code within that country for your SkypeIn number; you will have to choose from a predefined list of numbers for the digits that follow the area code. By choosing the country and area code carefully, you can have a SkypeIn number that best meets your needs—for example, to allow toll-free local calls for friends and family. Indeed, following this logic, you can pick and choose several SkypeIn numbers to minimize the hassle and cost of calls for your most frequent callers. This way, for example, you can have a local dial-in number for your home in New York and a dial-in number in Seattle for the convenience of your mom and dad living in the Seattle area.

Getting SkypeIn

The easiest way to get a SkypeIn dial-in number is to click on the **My Account** tab (which is the tab on the far right) of the Notification Bar in the Skype softphone. This displays a group of links, one of which is for SkypeIn, as shown in the following figure.

Signing up for SkypeIn.

Click on the link for **SkypeIn.** This opens a browser window and presents you with a form to sign in to your Skype account, which requires the same username and password that you use to sign in to the Skype softphone. Next, you are presented with a web page giving an overview of your account.

In your account overview page, click on the link **Buy SkypeIn Number.** This displays a page from which you can choose a country for your dial-in number. After you select a country, you need to choose an area code prefix. You are then presented with predefined numbers for that area code. Search among the available codes until you find one that strikes your fancy, and click on it. Choose a subscription period: 3 months costs 10 € ($12), and 12 months costs 30 € ($36).

Next, click on the **Buy selected number** button, which takes you to Skype's payment page. Follow the prompts to purchase your number.

After payment is complete, go to your Skype softphone and again click on the **My Account** tab (which is the tab on the far right) of the notification bar. You should see something like what's

> **Something Worth Knowing**
>
> Skype Voicemail is bundled with SkypeIn, so by becoming a SkypeIn subscriber you will also get voicemail for free. For details on using Skype Voicemail, see Chapter 7.

shown in the next figure, indicating that your newly minted dial-in number and voicemail account have been activated. Tell your friends, family, and colleagues what your new number is, and you're ready to receive calls from regular and mobile phones. If you want additional SkypeIn dial-in numbers, just repeat the procedure described in this section.

SkypeIn (and voicemail) activated.

The Least You Need to Know

◆ SkypeIn is a fee-based subscription service that allows you to receive calls to your Skype softphone from regular and mobile phones. That is, it gives you a regular dial-in number.

◆ You can obtain SkypeIn numbers not just for your local area code, but for other area codes, and indeed for other countries.

◆ You can have up to 10 SkypeIn dial-in numbers, all of which will ring at your Skype softphone.

◆ SkypeIn comes bundled with Skype Voicemail, so you won't miss calls even when you are away from your computer.

Voicemail and Call Forwarding

In This Chapter

◆ Learning about Skype Voicemail and call forwarding

◆ Signing up for Skype Voicemail

◆ Setting up and using voicemail

◆ Setting up and using call forwarding

Skype Voicemail and call forwarding make it so you never have to miss a call. These Skype features can be used separately or in combination. This chapter shows you how.

Skype Voicemail is a fee-based service that records your incoming calls when you are unable to take them yourself. At a later time and at your convenience, you can listen to messages from missed calls using the Skype softphone. An important plus is that Skype Voicemail works even when you're not running or signed in to the Skype softphone. So, even when your PC is powered off, Skype Voicemail records your missed calls.

> **Something Worth Knowing**
>
> Skype Voicemail is bundled with SkypeIn (see Chapter 6), so by becoming a SkypeIn subscriber you automatically qualify for voicemail.

Skype's call forwarding feature forwards calls to alternative Skype usernames and, if you are a SkypeOut subscriber, to regular or mobile phones. Call forwarding can forward calls to up to three alternative destinations. Note that in the case of calls forwarded to regular and mobile phones, you have to be a SkypeOut subscriber (see Chapter 5) and pay the appropriate per-minute call rate for the destination phone. Like Skype Voicemail, call forwarding even works when your PC is powered off; that is, calls are forwarded even if you are not signed in to the Skype account for which call forwarding has been set up. However, unlike Skype Voicemail, call forwarding is a free feature of Skype!

With Skype Voicemail and call forwarding, you need not miss another call again!

Getting Skype Voicemail

The easiest way to get Skype Voicemail if you are not a SkypeIn subscriber is to click on the **My Account** tab (which is the tab on the far right, and might have a SkypeOut credit balance amount on it if you are already a SkypeOut subscriber) of the Notification Bar in the Skype softphone. This displays a group of links, one of which is for Skype Voicemail, as shown in the following figure. Click on the link for **Skype Voicemail.** You will be asked to sign in to your Skype account, which requires the same username and password that you use to sign in to the Skype softphone. After signing in, you'll see the **Buy Skype Voicemail** web page.

Signing up for Skype Voicemail.

In the **Buy Skype Voicemail** web page, you will have two choices:

- Subscribe to voicemail for 3 months, at a cost of 5 € ($6).

- Subscribe to voicemail for 12 months, at a cost of 15 € ($18).

Choose a subscription period, 3
months or 12 months, and then
click on the **Buy this** button for
your choice and complete the payment process.

 Caution! _____

Skype Voicemail mes-
sages are limited to 10
minutes each.

Setting Up and Using Voicemail

You can personalize your voicemail message. You can also set up how
voicemail will respond to incoming calls.

Adding a Personalized Message to Your Voicemail

By default, you are given a rather
bland "The person you are trying
to call is not available ..." prerecorded message. This is what callers hear when they are redirected
to your voicemail. However, you
can replace the prerecorded message with a personalized greeting of
up to sixty seconds in duration.

 Something to Try _____

Being a Skype Voicemail
subscriber allows you to
send voice messages to
other Skype users, even
if they themselves are not
voicemail subscribers. To send
a voice message to another
Skype user, right-click on a
Skype user in your contacts
list, and from the popup menu
that appears, choose **Send
Voicemail**.

Go to **Skype: Tools: Options ...,** and in the window that appears,
click on **Call Forwarding & Voicemail.** A settings window appears, as
shown in the next figure.

Look under the heading "Welcome Message" at the bottom of this
page. To record a personalized greeting for your voicemail, click on the
middle square button with the red circle on it. Speak into your microphone (or headset or handset) to record your greeting and then either
click on the same red button again to cut the recording short, or let the
recording run to the end of its allotted 60 seconds. You can now review
your greeting by clicking on the green square button with an arrow on
it. Now, when callers are directed to your voicemail, they will hear your
personalized greeting message.

Skype Voicemail and Call
Forwarding settings.

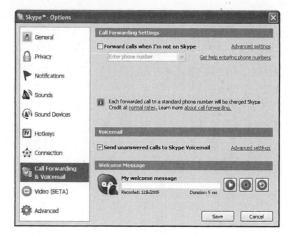

If you're not satisfied with the greeting, you can either rerecord it by
hitting the middle red square button again and repeating the recording procedure, or you can click on the red square button with a circular
arrow on it to reset your greeting to the default message.

Directing Calls to Voicemail

To have unanswered incoming calls directed to your voicemail, simply put a tick in the box opposite Send Unanswered Calls to Skype
Voicemail (see the preceding figure). If you want to control how and
when calls are redirected to voicemail, click on the **Advanced Settings**
link alongside Send Unanswered Calls to Skype Voicemail. This gener-
ates a window in which you can customize how your voicemail behaves,
as shown in the following figure.

Skype Voicemail advanced
settings.

Listening to Voicemails

Missed calls, like other Skype events, are flagged for your attention in the New Events tab (which is the second tab from the right) of the notification bar. Click on this tab to see recent missed events, including voicemail messages that have been left for you, as shown in the following figure.

Skype events, including voicemails.

Clicking on a voicemail event in the New Events tab of the notification bar takes you to the History tab, as shown in the next figure. Voicemails that you have not yet listened to are flagged by a red dot with a star. By clicking on the green round button with an arrow on it to the right of the voicemail item, you can listen to your voicemail using the audio-out device you have configured for Skype (for example, headphones or speakers).

A Skype Voicemail event.

You can listen to unheard voicemail messages on any computer in which you are signed on to Skype. So even if you are traveling, you can listen to your latest voicemails as long as you have access to a computer and an Internet connection.

Something to Try

You can use the pull-down menu at the far left of the address bar to filter the history list to display only missed calls, incoming calls, outgoing calls, all calls (incoming and outgoing), voicemails, transferred files, or chats. If the address bar is not visible, you can display it by going to **Skype: View: View Address bar**. You can clear a single item from the history tab by right-clicking on it and, from the popup menu that appears, clicking on **Delete This Entry**. To clear all events from your history, go to **Skype: Tools: Clear History** on the Main Menu.

Setting Up and Using Call Forwarding

Call forwarding allows you to reroute incoming calls to other phone numbers, including Skype numbers, regular numbers, and mobile numbers. To forward calls to regular and mobile phones, you must be a subscriber to SkypeOut, and forwarded calls will be billed just as though you had personally called the forwarding destination phone from your PC using Skype. Incoming calls to your Skype account from both other Skype users and from regular and mobile phones (if you're a SkypeIn subscriber) are forwarded in the same way.

Setting Up Call Forwarding

To set up call forwarding, go to **Skype: Tools: Options ...**, and in the window that appears, click on **Call Forwarding & Voicemail.** This displays the option settings for both call forwarding and voicemail. To activate call forwarding, check the box opposite **Forward calls when I'm not on Skype**.

The default setting, also called the basic setting for call forwarding, allows you to forward a call to one destination. Advanced call forwarding allows you to forward incoming calls to up to three different destinations. When a call is forwarded, it rings all forwarded destinations at the same time; whichever destination picks up first gets the call.

You can switch between basic and advanced call forwarding settings by clicking on the links **Basic Settings** and **Advanced Settings,** respectively.

Skype call forwarding settings.

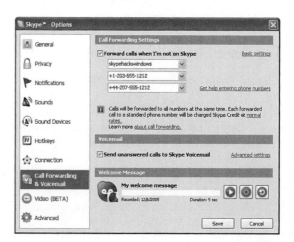

Using Call Forwarding

How quickly calls are forwarded depends on your online status. The following table indicates the how quickly calls are forwarded based on your online status.

Call Forwarding Behavior Based on Online Status

Online Status	Call Forwarded
Skype not running	Immediately
Online	Between three and five rings
Offline	Immediately
Skype-Me	Between three and five rings
Away	Between three and five rings
Not available	Between three and five rings
Do not disturb	Between three and five rings
Invisible	Between three and five rings

Armed with the information in this table, you can decide how best to use call forwarding. You can set your online status (Online, Skype-Me, Away, Not available, or Invisible) so that if you are at your PC you have the option of taking the call then and there; and only if you don't pick up will Skype forward the call. Or you can set your online status (to Skype not running or Offline) so that Skype always and immediately forwards all incoming calls to the destination (or destinations) of your choice.

The Least You Need to Know

◆ By using Skype Voicemail and call forwarding you never have to miss an incoming call.

◆ Skype Voicemail is a fee-based subscription service that records messages from missed calls.

◆ You can listen to voicemails from wherever you are signed into Skype.

◆ Call forwarding allows you to forward incoming calls to up to three other Skype users or regular phones. Call forwarding to other Skype usernames is free.

◆ Using your online status, you can control call forwarding so that it forwards all incoming calls immediately or gives you the option of taking calls if you happen to be close by your PC.

Skype Zones

In This Chapter

◆ Learning about Skype Zones

◆ Signing up for Skype Zones

◆ Using Skype Zones

Skype Zones is a fee-based service that enables you to use Skype while traveling. By becoming a Skype Zones subscriber, you gain access to more than 18,000 Skype-friendly wireless Internet access points dotted around the globe. Skype Zones is a service provided in partnership with Boingo Wireless (www.boingo.com).

Skype Zones Options

There are two pricing options for the Skype Zones service:

◆ **Skype Zones Unlimited** gives you "unlimited" access to the Skype Zones wireless network for a monthly fee of $7.95.

◆ **Skype Zones AsYouGo** gives you a single connection that you can use for up to two consecutive hours for $2.95.

Clearly, if you move around a lot and don't typically use Skype in blocks of two hours at a time, then the Unlimited plan is for you. On the other hand, if you only occasionally use Skype when on the move, the AsYouGo plan might be more appropriate.

Caution! _____

If you read the fine print of the Skype Zones agreement you will find that their "Unlimited" plan isn't really unlimited. Instead, it allows you to connect to the Skype Zones network for a maximum of 300 minutes per month.

Becoming a Skype Zones subscriber allows you to use all the features of Skype, but it doesn't give you general access to the Internet. For access to the Internet for web browsing, e-mail, and so on, you will have to use a different service.

Checking to See if You Are Wireless Enabled

Before launching yourself into this chapter any farther, you should first understand that Skype Zones is a wireless network service, so your PC must be wireless enabled in order to make use of it. To check whether your PC has a wireless network adapter installed, go to **Start: Control Panel** and double-click on **Network Connections.** This should open a window that displays the available network connections for your PC. If, under the category LAN or High-Speed Internet, you have an entry named Wireless Network Connection (or something similar), you're good to go for this chapter. If not, you'll need to install and set up a wireless network adapter for your PC before you can use Skype Zones.

Signing Up for Skype Zones

To sign up for Skype Zones, complete the following steps:

1. Go to www.skype.com and click on the **Skype Zones** link, which can be found towards the bottom of the home page.

2. From the Skype Zones web page that is displayed, click on the link **Sign up for Skype Zones now to get started.**

3. Next choose the plan you want, Unlimited or AsYouGo, put an acceptance tick in the box underneath the service agreement, and click **Next.**

4. Enter your personal details (name, address, and so on) into the web form that appears and click **Next.**

5. Because the Skype Zones service is provided by Boingo, you must next create a Skype Zones user account (a user name and password, which need not be the same as your Skype user name and password) with Boingo. Enter your desired user name and password for Skype Zones, and then click **Next.**

6. Enter your credit card information for billing purposes, and click **Next.**

7. Finally, review all the information you have entered so far. If the information is correct, click **Submit.**

Phew! Now you are a Skype Zones subscriber.

However, there is one more thing to do before you can actually use Skype Zones: you must download and install the Skype Zones software.

After clicking on the Submit button on the final page of the signup process, you will be presented with a web page from which you can download the Skype Zones software. Just click on **Download Skype Zones For Windows** and in the popup window that appears, click on the **Run** button. Follow the instructions of the Skype Zones installation program, which will install and then run the Skype Zones program, which is shown in the following figure.

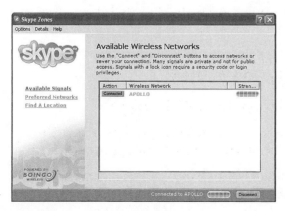

The Skype Zones client program.

Using Skype Zones

To use Skype Zones, you must first find an access point. Then, once you are within wireless range of an access point, you must connect to it. Fortunately, Skype has made both tasks relatively easy.

To find a Skype Zones access point, you have two choices. You can use either a web browser or the Skype Zones client program.

Finding a Skype Zones Access Point Using a Web Browser

To find an access point using a web browser, first go to www.skype. com. Click on the **Skype Zones** link, which can be found toward the bottom of the home page. On the Skype Zones web page that appears next, clicking on the link **Find Skype Zones** will display the web page shown in the next figure.

Web page search for Skype Zones access points.

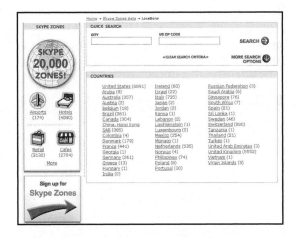

Suppose you'll be in Madrid, Spain, some time during the next few days and you want to find out where you can access Skype Zones. Type "Madrid" in the city search criteria, and then click on **Search.** You will be given the list of Skype Zones available in Madrid, as shown in the next figure.

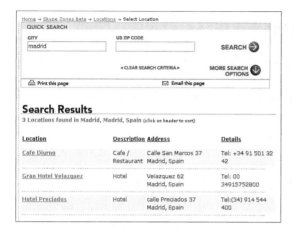

Skype Zones access points for Madrid, Spain.

Finding a Skype Zones Access Point Using Skype Zones Client Program

To use the Skype Zones client program to search for access points, first open the client by going to **Start: All Programs: Skype Zones: Skype Zones.** Next, click on the link **Find A Location,** which displays a search page, as shown in the next figure.

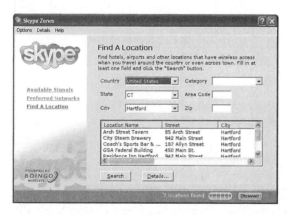

Searching for access points using the Skype Zones client program.

Frankly, the only advantage of using the search facility in the Skype Zones client program is that it works even when you are not connected to the Internet (the list of available access points is downloaded to your PC and is periodically updated). However, it doesn't always have the most up-to-date list of available access points.

Connecting To and Using a Skype Zones Access Point

Once you've located an access point, it's time to connect to Skype Zones.

First, make sure you are running the Skype Zones client program (**Start: All Programs: Skype Zones: Skype Zones**). When you run Skype Zones and are in range of a Skype Zones access point, a window should pop up that invites you to sign in to the Skype Zones network. Alternatively, you can open the Skype Zones client (right-click on the **Skype Zones** icon in the Windows system tray and choose **Open Skype Zones**), and then click on the link **Available Signals.** This should display the Available Wireless Networks page in the Skype Zones client. To connect to a specific access point, simply click on the blue **Connect** button opposite the desired wireless network name. To disconnect, just click on the green **Connected** button. If the wireless access point is part of the Skype Zones network, you will be prompted to sign in when you connect. Once connected, your online status in the Skype softphone will change to Online and you can start using Skype as you would from any regular Internet connection.

> **Something Worth Knowing**
>
> You can cancel the Skype Zones Unlimited plan by e-mailing zonessupport@ skype.net or calling Boingo at +1-866-526-4646 (+1-310-460-3906 outside the United States) five days before your next billing date.

The Least You Need to Know

- ◆ Skype Zones, a fee-based subscription service, enables you to use Skype from any of the more than 18,000 wireless access points around the world.

- ◆ To use Skype Zones you must download and install a client program that manages your connections to the Skype Zones wireless network.

- ◆ You can search for the locations of Skype Zones access points using either a web browser or the Skype Zones client program.

Chapter 9

Maximizing Your Savings Using Skype

In This Chapter

- ◆ Saving money using Skype
- ◆ Estimating your potential savings
- ◆ Finding even cheaper alternatives to Skype
- ◆ Canceling services and getting a refund
- ◆ Paying taxes

Your regular and mobile phone bill is normally made up of two parts: cost of calls, and fixed charges and taxes. You can use Skype to reduce the overall cost of your phone calls, and reduce—in some cases even eliminate—your fixed charges and taxes.

The cost of an individual call is the duration of the call, measured in minutes (and sometimes fractions thereof), multiplied by the call rate. For example, a 7 minute call at a call rate of 5 cents per minute will cost you 35 cents. The lower the call rate for a given destination, the lower your cost of calls to that destination. You can use Skype's very low per minute call rates to save money on your overall cost of calls.

Over the years, most people have seen their phone bills accumulate lots of fixed charges and taxes. Indeed, before I switched to Skype, fixed charges and taxes amounted to 56 percent of my regular telephone bill! Some taxes relate to your cost of calls and so decline as your cost of calls decrease. But most are truly fixed, in the sense that even if you don't make a single call during any given month, you still have to pay them. To eliminate the fixed charges and taxes, you'd need to reduce the number of phone lines entering your home or eliminate them altogether. By using Skype you can do this and so avoid many fixed charges and taxes—and it's all legal!

In this chapter I describe ways in which you can use Skype to save money—in many cases a lot of money—off of your current phone bill.

Estimating Your Potential Savings

Estimating your potential savings from using Skype for some or all of your phone calls requires nothing more than a pencil, a notepad, and just a little basic math.

Savings will vary. It's even possible that some people might not save any money at all by switching to Skype, though I can't imagine how that might come about. If you're very aggressive in switching to Skype, you can save an awful lot of money. For instance, I shaved 82 percent off my old annual phone bill, which means I save $679 per year!

Estimating Savings on Your Cost of Calls

Look at your most recent phone bill, or bills, as maybe you have both a regular phone and mobile phone. Or better yet, if you still have them, gather several months' worth of bills. From your itemized list of calls, pick out the 10 most expensive calls.

Now go to the Skype Dialing Wizard at www.skype.com/products/skypeout/rates/dialing.html, and for each call you have chosen, enter the telephone number and note down the per-minute call rate that the Skype Dialing Wizard gives you. An example of how to get the per minute call rate for a specific number, using the Skype Dialing Wizard, is as shown in the following figure.

SkypeOut Dialing

Need some help figuring out how to dial a number on SkypeOut? This page is for you, then. Whether you're calling your nextdoor neighbour or a friend living in another country you always need to dial the full international number.

Currency for SkypeOut rates: United States, Dollars ▼

1 **What phone number are you dialling?**
Enter the full area code and phone number just as you would when dialing with a regular phone.

+1-202-456-1111

2 **What country or region is the number in?**
Select the country or region below.

United States ▼

3 **Dial the number like this in Skype**
Always dial a full international number in Skype.

+1 2024561111 *Dial number in Skype*
└ The telephone number
└ International dialing for United States

4 **Rates for dialing this number**
Rates listed here are for reference only. View your call list after your call for exact charges.

$ 0.021 per minute

You can use the Skype Dialing Wizard to find out the call rate for specific numbers.

Next, round the call duration of each of your chosen calls so that they're expressed in whole minutes. For example, round a 45-second call up to 1 minute, and round a 5-minute, 25-second call up to 6 minutes. Now multiply these whole-minute call durations by the corresponding SkypeOut call rates you obtained from the Skype Dialing Wizard. Compare these totals with the actual cost of each of the calls on your phone bill. If your calculated numbers are less than your actual numbers from your phone bill, you would have saved money using Skype. And, provided your calling habits don't change much, you will most likely save money if you use Skype in the future. It's as simple as that.

Something to Try

In the Skype Dialing Wizard, you can change the display currency from euros (the default) to U.S. dollars or any other currency by using the pull-down list of currencies at the top of the web page.

> **Caution!** _____
>
> Keep in mind that there's no concept of local, long distance, inter-
> state, or international calling in Skype. All call rates are global rates,
> in the sense that a SkypeOut call rate for a particular destination is good
> when dialing from anywhere on the planet.
>
> If you make a lot of local, toll-free calls—calls that are not itemized on
> your regular phone bill—then making these calls using Skype will cost you
> money. Toll-free local calls become toll calls when dialed from Skype. This
> should be factored into your cost-saving calculations. To estimate your cost
> of local calls if you were to fully switch to Skype, estimate your monthly
> minutes of local calling and multiply that number by the SkypeOut rate for
> your locality (country code plus area code).
>
> There's nothing stopping you from keeping a regular phone for local calls
> and switching only your long distance and international calls to Skype.

Estimating Savings on Your Fixed Charges and Taxes

The only really meaningful way to attack your fixed charges and taxes
is to eliminate one or more phone lines. For example, suppose you have
two telephone lines, and your monthly fixed charges and taxes are $50.
If you eliminate one telephone line and replace the other with Skype,
you will save about $25 a month on fixed costs alone.

The bottom line is this: you can't start attacking fixed charges and
taxes unless you can shed phone lines. If you are able to eliminate one
or more lines, sum all the monthly fixed charges and taxes on your
phone bill, and then divide that sum by the number of phone lines you
have. This will give you the fixed charges and taxes per phone line per
month. For each phone line you can eliminate, you will save about that
much money, every month.

> **Caution!** _____
>
> There are good reasons
> for keeping at least one
> regular telephone line; not least
> among them is to have access
> to 911 emergency services,
> which Skype does not sup-
> port. This topic is covered in
> Chapter 20.

More Money-Saving Strategies

In addition to reducing the per-minute cost of your calls and reducing or eliminating fixed charges and taxes, there are other money-saving strategies you can employ by using Skype. I'll skip the fatuous advice, such as "make fewer calls," or "use e-mail instead," because this chapter is about saving money by using Skype, or alternatives to Skype. Though I must say, "make fewer calls" has worked wonders in my family, and was accompanied by a family-wide sigh of relief!

Make More Free Calls

Remember that you can call anyone who uses Skype for free. That means that if you convince people you call frequently, such as family members, friends, and even business contacts, to join Skype, you could save a lot of money. For example, suppose you spent $50 last year speaking to Uncle Tom on a land line. This year, however, you both decide to join Skype. That means you'll save $50 this coming year, and each and every year beyond that. Indeed, you needn't watch the clock any longer, and so can speak with Uncle Tom more often and for longer. In short, the more people you convert to Skype, the more money you will save.

Make Calls Cheaper for People Who Call You

Using Skype, you can even reduce the regular phone bills of others. You can do this by subscribing to one or more SkypeIn numbers. (See Chapter 6 for a detailed discussion of SkypeIn.)

How you do this is best illustrated using an example. Suppose you live in Baltimore, but have a lot of friends and family in London, England. By choosing a London number for SkypeIn, those same friends and family can call you in Baltimore for the cost of a local call—instead of an international call. Indeed, you can have several dial-in numbers (up to 10) dotted around the globe. Friends and family anywhere can then call what is a local number for them at their local rate (sometimes for free), and these calls will then be routed for free (courtesy of the Skype

P2P network) to wherever you happen to be signed in to Skype. This way, friends and family can save money by calling your SkypeIn numbers at their local rates, while you can receive such calls anywhere on the planet!

Of course, this strategy will save money for others but not for you, as you will have to pay to subscribe to SkypeIn numbers. But think of all the goodwill it will buy you!

Alternatives to Skype

Skype isn't the only game in town! Other Internet telephony providers are even cheaper than Skype for some types of calls. You can mix and match alternatives to Skype to get the overall best deal.

Don't be afraid to shop around for the best deal. Using the tips that follow, and by searching using an Internet search engine such as Google (using search terms such as "VoIP" and "softphone"), you will quickly find plenty—and in some cases cheaper—alternatives to Skype.

However, a few words of caution. When considering alternatives to Skype, be sure to consider issues such as voice quality (Skype's voice quality is very good), privacy, security, ease of use, support for advanced features such as video and chat, and software and hardware compatibility with your PC.

To get you started, here are some alternatives to Skype:

♦ **www.voipstunt.com** is a softphone that you install on your PC which allows you to make free calls to other VoIPStunt users. In addition, it allows you to make free calls to regular landline phones in several countries, including the United States, Canada, United Kingdom, Australia, and New Zealand. (See the VoIPStunt website for a complete list.) Even more amazing is the fact that you can also call mobile phones for free in some of these countries, including the United States and China. More amazing still is that you can get a free regular dial-in number, although the list of countries for which you can get them is more restricted.

◆ **www.voipbuster.com** offers free calls to regular landline phone numbers in the following countries: Andorra, Australia, Austria, Belgium, Bulgaria, Canada, Chile, Colombia, Croatia, Cyprus, Denmark, Estonia, Finland, France, Georgia, Greece, Hong Kong, Iceland, Ireland, Italy, Japan, Latvia, Liechtenstein, Luxembourg, Malaysia, Monaco, Mongolia, Netherlands, New Zealand, Norway, Panama, Peru, Portugal, Puerto Rico, Singapore, Slovenia, South Korea, Spain, Sweden, Switzerland, Taiwan, Thailand, and Venezuela. Like VoIPStunt, VoIPBuster offers free regular dial-in numbers for some countries. Check the VoIPBuster website for details and restrictions.

◆ **www.gizmoproject.com** is a softphone with features and services that are very similar to Skype.

◆ **www.freeworlddialup.com** is a service that is based on open standards and so can work with different softphones. It offers free softphone-to-softphone calls, and free calls to some regular phones.

◆ **www.google.com/talk** is a softphone from the folks at Google. It enables you to speak for free with other Google Talk users, and supports instant message chat. Unlike Skype, you can use Google Talk to chat with other chat message client programs.

◆ **messenger.yahoo.com** is an instant messaging chat client that also supports free PC-to-PC voice calls. It also offers low per-minute rates for calls to regular phones.

◆ **messenger.msn.com** is a softphone that supports voice calls, video, and chat messaging.

Canceling Services and Getting a Refund

Skype makes it easy to cancel its services. No long contracts, no exit fees—in short, no lock in! Indeed, for some Skype services, if you cancel, you can get a refund—either in whole or in part—for the money you spent. Try that with a regular or mobile phone company!

> **Caution!**
>
> Remember that SkypeOut credits expire if you don't make any SkypeOut calls during a 180-day period. Make at least one SkypeOut call every 180 days or less and your credits are safe.

Canceling Services

To cancel SkypeOut, SkypeIn, or Skype Voicemail service, you need simply submit a support request. Go to support.skype.com and click on the link **Submit Support Request.**

To cancel Skype Zones service, e-mail zonessupport@skype.net or call Boingo at 1-866-526-4646 (1-310-460-3906 outside the United States) at least five days before your next billing date.

Getting a Refund

You can cancel SkypeOut, SkypeIn, or Skype Voicemail at any time and get a refund. Here's what you do for each service:

◆ **SkypeOut:** If you have spent less than 1 € ($1.20) of your last subscription payment, Skype will refund the whole amount of that payment. If you spent more than 1 €, Skype will refund only the remaining balance.

◆ **SkypeIn and Skype Voicemail:** If you ask for a refund within 30 days of your last subscription payment, Skype will refund the whole amount of that payment. After more than 30 days, Skype will refund only the amount remaining for your subscription on a pro-rata basis.

> **Something Worth Knowing**
>
> If you bought your Skype subscription services or SkypeOut credits using vouchers, you do not qualify for a refund. Also, a user of Skype services allocated by a Skype Control Panel (see www.skype.biz) administrator cannot get a refund of any services; only the administrator can do that.

Refunds take five to seven business days to appear on your online financial statements.

You can't get a refund for Skype Zones, but there's no long-term contract or exit fee either!

Paying Taxes

If you use only the free features of Skype, you need not pay any taxes. Period.

However, if you are a subscriber to one or more of Skype's fee-based services, you might be obligated to pay taxes. Subscribers whose billing address is in a country that is a member of the European Union will be charged Value Added Tax (VAT) when they purchase services. But subscribers elsewhere, including the United States, will not be charged any taxes. This, however, does not necessarily mean you don't owe taxes on payments for those services!

In the United States a form of tax called the Sales & Use tax is levied at the state level, and sometimes the local level, on goods and services you purchase outside of your home state. If such a tax is not collected by the seller, then you—the buyer—are required by law to pay it. This is a complex issue beyond the scope of this book, and so you should check with your local tax authority or seek professional tax advice regarding the rules for making Skype purchases.

If you subscribe to Skype's fee-based services and live somewhere other than the European Union or the United States, you will need to determine the tax status of Skype purchases for yourself by checking with the tax authorities for your jurisdiction.

The Least You Need to Know

◆ Your regular phone bill can be divided into two parts: cost of calls, and fixed charges and taxes. Skype has the potential to save you money on both parts.

◆ Figuring out your potential savings from switching to Skype is simple, and requires only an old phone bill, pencil, paper, and just a little math.

◆ There are alternatives to Skype when making calls over the Internet. Some are even cheaper than Skype for some types of calls.

◆ You can cancel Skype's fee-based subscription services at any time. Often, you'll even receive a refund.

◆ Depending on where you live, you might have to pay taxes on your purchases of fee-based Skype services. If in doubt, seek professional financial advice.

Part 3

Customize and Extend Skype

"Customize, personalize, and extend" is the mantra for this section. Although the Skype software already comes bundled with a lot of neat features, there are a number of things you can do to customize it to better suit your needs and also improve Skype as a productivity tool. In addition, you can extend Skype in weird and wonderful ways using third-party add-ons and so further bend Skype to your way of doing things.

Chapter 10

Configuring Skype to Meet Your Needs

In This Chapter

- ◆ Making Skype work the way you want it to
- ◆ Configuring your online profile and online status
- ◆ Changing the look of Skype's user interface
- ◆ Contact management
- ◆ Configuring Skype's options

Although it is impossible for a widely used application such as Skype to be all things to all people, there are things you can do to bend it to your way of doing things. By customizing Skype to better fit your needs, not only do you make Skype more fun to use, you also make it more efficient.

In this chapter you will find ways to customize such things as your online profile and online status, your contacts list, the look of Skype's softphone, and other options that control the behavior of Skype's softphone.

Not covered in this chapter is configuration and use of Skype Video, or Voicemail and call forwarding, as these topics were covered in Chapters 4 and 7, respectively. Also not covered in this chapter are privacy and security features, because these topics are sufficiently important to warrant a section of their own—see Part 4 of this book.

For users with disabilities or other special needs, Part 6 of this book expands on this chapter's customization to address the issues of accessibility and usability. But Part 6 is also for people without disabilities who simply want even more control over how the Skype softphone looks and behaves.

Your Online Profile and Online Status

Your online profile is your public face to the Skype community. It also provides details that others in the Skype community can use to search for and find people. The fewer details you provide, the less visible you become in the Skype community. Conversely, the more details you provide, the higher your visibility in the Skype community. Even though you can change your online profile anytime you want, it makes sense to put some thought into how visible you want to be from the start.

Another way of controlling your visibility and availability within the Skype community is to change your online status. Your online status determines how contactable you are.

Something Worth Knowing

The details you enter in your Skype profile need not be accurate. Indeed, you can simply make them up if you want. One reason you might choose a fictitious persona is to avoid unwanted calls from strangers. Skype, like any online community, has its undesirable elements. (Though it must be said, they make up a tiny fraction of the online community.) So if you are a woman who wants to attract less attention from the opposite sex, you might want to set your gender to male. It is likewise prudent to choose a Skype name that doesn't attract unwanted attention; for example, a Skype name that includes the words "hot" and "babe" is likely to get you a lot of attention!

Online Profile

To change the details of your online profile at any time, go to **Skype: File: Edit My Profile ....** This displays the My Profile form in a window, as shown in the following figure.

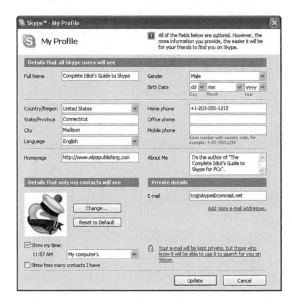

Skype online profile.

The details of your online profile are divided into three categories:

- ◆ Details that all Skype users will see

- ◆ Details that only your contacts will see

- ◆ Private details

Most of the fields in the My Profile form are self explanatory, but there are a couple of things worthy of note.

First, you can add a picture to your profile. When you call or chat with another Skype user, this picture appears in his Skype softphone. Skype comes with a set of default pictures that you can use, or you can purchase more using the Personalise Skype service (see Chapter 11). To change the picture for your profile, simply click on the **Change ...** button and a window titled My Pictures will pop up with a list of available pictures. Click on a picture and then click the OK button to change the picture in your online profile. You can find more information on jazzing up your online profile with pictures in Chapter 11.

Second, including a valid e-mail address in your profile has both advantages and disadvantages. A valid e-mail address is necessary in order to reset your password should you forget your current password (see Chapter 15), and it also means that you will get 30-day and 7-day warnings when your SkypeOut credits are in danger of expiring. On the other hand, adding an e-mail address to your online profile also enables others to search for you using this e-mail address; that is, a Skype search using your e-mail address will return your Skype user name. This means that friends who know your e-mail address, but not your Skype user name, can find you. But it also means that people who you don't want to talk with can find your Skype user name if they know your e-mail address.

Online Status

Before discussing how to change your online status, you should know what each status means:

- **Offline**   The Skype user is offline. Being Offline means that the user has disconnected from the Skype network.

- **Online**   The Skype user is online. Online means that the Skype user is ready to make and receive calls, chat, and voicemail. It also means that the user can send and receive files.

- **Skype-Me**   The Skype user is in Skype-Me mode. Skype-Me mode means, in effect, that you are making yourself available to the wider Skype community, because while in this mode your privacy settings are ignored. Anyone can call you while you are in Skype-Me mode.

- **Away**   The Skype user is away from Skype. Online status changes to away when the PC on which Skype is running has been inactive for a period of time that you can specify (see Chapter 15); the default is five minutes.

- **Not Available**   The Skype user is not available. This online status is similar to Away, but indicates that the user has stepped away

from the PC running Skype for a prolonged period. You can specify the period of inactivity before Skype switches to Not Available
(see Chapter 15); the default is 20 minutes.

- **Do Not Disturb**   The Skype user does not want to be disturbed.
  While in this mode, you won't be pestered by popup notifications
  of incoming calls or chat.

- **Invisible**   The Skype user has chosen not to make his or her
  online status visible to others. However, when online the privacy
  settings for that Skype user remain in effect.

An illustrated list of all Skype icons can be found in Appendix D.

To change your online status, go to **Skype: File: Change My Online
Status,** and then choose the online status you want. Or you can use the
pull-down menu at the bottom left of Skype's softphone main window.

Change the Look of Skype's User Interface

In terms of changing the look, and in some ways the feel, of Skype's
graphical user interface, you have the following two options:

- Change how Microsoft Windows displays the graphical elements
  of all application windows. Clearly, these changes are global, in
  that they change the look of your entire desktop.

- Selectively display, or not display, some graphical elements of
  Skype's softphone main window. These changes affect only Skype's
  window, leaving the rest of your desktop unchanged.

The following sections walk you through each of these options.

Changing the Look of the Windows Desktop

To change the look and feel of your entire Windows desktop, including the Skype softphone main window, right-click on any part of your
Windows desktop that has no icons; from the popup menu that appears,
click on **Properties.** This displays the **Display Properties** window, as
shown in the next figure.

A high-contrast and enlarged font version of the Skype softphone with the Windows desktop Display Properties window beside it.

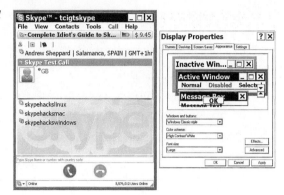

You can change the look and feel of the Windows desktop using either the Themes tab or the Appearance tab.

Themes are predefined settings that establish the look and feel of all graphical elements for your Windows desktop. However, using only the Themes tab you can't tweak individual elements of your desktop—for example, the font size for menu names.

Clicking on the **Appearance** tab, on the other hand, gives you a lot of control over the look and feel of most individual graphical elements (buttons, menus, colors, and so forth) of your Windows desktop. If you can find a Windows theme (many are available for free on the Internet) that meets your needs, then fine. Otherwise, you might want to play with the controls available through the Appearance tab of the Display Properties window.

Changing the Look of Skype's Softphone

You can change the look of Skype's softphone by selectively displaying, or not displaying, elements of its user interface. To switch the display of elements on and off, go to **Skype: View**, check the elements you want to be displayed, and remove checks from elements you do not want to be displayed. You can experiment with this until Skype looks the way you want it to. A minimalist's idea of the Skype softphone might resemble the following figure.

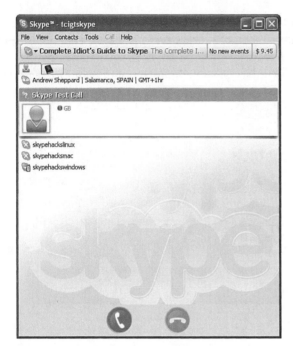

A minimalist's Skype soft-phone.

Contact Management

Your contact list is a tiny subset of the Skype online community with whom you communicate more often. If you are a SkypeOut subscriber, you can add SkypeOut contacts as well. Contact names are displayed in your Skype softphone for convenience and ease of use. You can also call or chat with someone not in your contact list by using her Skype user name or phone number by manually entering in the address bar her name or number each time you want to communicate with her. However, organizing and managing your contacts list can be a real productivity boost!

> **Something Worth Knowing**
>
> Your contact list is stored on Skype's network, so it will be available wherever you sign on to Skype.

Import Contacts

Skype is unlikely to be the first application you run on your PC that maintains a list of contacts. So to help get you started, Skype has a tool for importing contacts that already exist on your computer. It will import, if you so choose, contacts into Skype from the following applications: Microsoft Outlook, Microsoft Outlook Express, Microsoft MSN Messenger, and Opera (www.opera.com). To import your existing contacts into Skype, go to **Skype: Contacts: Import Contacts …,** and follow the instructions of Skype's contact import wizard tool.

Search For, and Add, Contacts

To search for other Skype users to add to your contact list, either click on the Search icon on the Skype softphone toolbar, or go to **Skype: Contacts: Search for Skype Users ….** This displays the search window shown in the next figure.

Search for other Skype users.

Using the search window, you can filter the list of all users in the Skype online community down to only those people you are interested in and that you might want to add to your contacts list. Note that the users you are searching for don't need to be signed in to Skype for you to find them and add them to your contact list. When you find someone you want to add to your list, highlight her entry in the list and then click on the **Add Selected Contact** button. When you're finished adding new Skype contacts, click on the **Close** button.

An alternative way to search for, and add, new Skype contacts is to click on the **Add Contact** icon on the toolbar of the Skype softphone.

Alternatively, you can go to **Skype: Contacts: Add a Contact ....**
Either method will display the **Add a Contact** window, as shown in the
following figure.

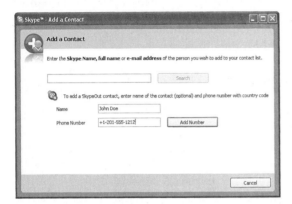

*Add a new contact, in this
case, a SkypeOut contact.*

If you are a SkypeOut subscriber, the preceding method is the only way
you can add regular and mobile phone contacts to your contacts list
without first making a call. In the **Add a Contact** window, just click
on the link **To add a SkypeOut contact, click here** (this link is only
displayed if you are a SkypeOut subscriber). Otherwise, when you hang
up after dialing a phone number not in your contacts list, you will be
invited by Skype to add that number to your list.

Using the **Add a Contact** window, you can search for a Skype user
by Skype user name, full name, or e-mail address. Entering a search
string and clicking the **Search** button (or the **Search Again** button
if you are conducting additional searches) presents you with a list of
matching Skype users. Again, simply highlight a name and click on the
Add Selected Contact button to add the name to your contacts list.
Alternatively, if you are a SkypeOut subscriber, clicking on the link **To
add a SkypeOut contact, click here** enables you to enter a name and a
regular phone number for that name.

By whatever method you add a new Skype contact to your contacts list,
when adding the contact, you will be given the opportunity to send
him a personal message, as shown in the next figure. In the interests of
everyone's privacy, just adding a contact to your list does not automatically give you access to that person's contact details (for example, that
person's online status). A person you invite to join your contact list can
choose to accept the invitation, ignore it, or block you from contacting

him ever again. If a contact accepts your invitation to share his contact details with you, his entry in your contacts list will provide more information; if not, all you will see next to his name is a gray icon with a question mark on it, and when you mouse over his entry in your contact list you will see a message saying, "This user has not shared his/her details with you".

Request that a Skype user added to your contact list share his or her contact details with you.

Rename Contacts

When you add a Skype user to your contacts list, the name that is displayed by default is the full name from the person's Skype profile, or if no such name exists, his or her Skype user name. If you are a SkypeOut subscriber, you are asked to name phone numbers as you add them to your contacts list. In either case, you can change the name of any contact to a name of your choice to help you better organize your contacts list.

To rename a contact, right-click on it in your contacts list. Then, from the popup list that appears, choose **Rename ...**, and type in a new name then press the enter key or click elsewhere with your mouse. The name will change and the entry will reposition itself in your contacts list based on alphabetical ordering (within the group to which it belongs, if you are using grouping, which is discussed later in this chapter).

Something to Try

In addition to naming contacts for your own convenience, you can also assign speed-dial numbers (ranging from 0 to 99) to your most popular contacts. After you've assigned a speed-dial number to a contact, you can call that contact by inserting the contact's speed-dial number in the address bar and then hitting the enter key on your keyboard or clicking on the call button. To assign a speed-dial number to a contact, select a contact in your contacts list by right-clicking on the contact with your mouse, and from the popup menu that appears, choose **Set Speed Dial.** In the window that opens, enter a speed-dial number and click on the OK button. That contact is now only a digit or two away!

Remove Contacts

To remove an entry from your contacts list, right-click on that entry, and from the popup menu that appears choose **Remove From Contacts.** It's as simple as that!

Caution!

Removing a contact from your contacts list does not block that user from contacting you. Blocking and unblocking contacts is covered in Chapter 15.

Group Contacts

You can group your contacts lists into various categories only if you have enabled **Show Contact Groups.** To enable contact grouping, go to **Skype: View** and put a check mark against **Show Contact Groups.** Contact grouping adds a group-management bar to the top of the panel displayed when you click on the **Contacts** tab, as shown in the following figure.

Using the contact groups bar you can create new groups and selectively display groups of contacts in your contacts list. To create a new group click on the small round button with a "+" on it, and to selectively display groups simply click on the small round button with a ">>" on it; both buttons can be found on the far right of the contact groups bar. In the case of the previous figure, groups named "People and Friends" and "Skype Testing" are used. However, you are free to choose how many groups you want, and how they are named.

Contact groups.

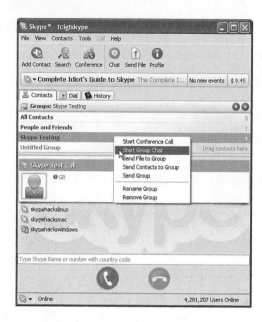

In addition to filtering and organizing your contacts list, another useful feature of contact groups is that you can carry out group actions on
them. By selecting a group and right-clicking on it, you can then use
the popup menu that appears to carry out actions for all the contacts in
that group (see the previous figure): Start Conference Call, Start Group
Chat, Send File to Group, Send Contacts to Group, and Send Group
which sends only those contacts in that group to a list of named Skype
users. Also, using this popup menu, you can Rename and Remove existing groups. In short, contact groups are a great way to organize your
contacts list and to carry out actions for a group that would otherwise
be tedious and cumbersome.

Hide Contacts

As a further aid to organizing your contacts list, you can shorten it by
choosing to hide users who are offline, or users who are not sharing
their contact details with you, or both kinds of users. To hide contacts,
go to **Skype: View: Hide Contacts That Are,** and put a check mark
against **Offline** or **Not Sharing Details.** This is a simple, quick, and
effective way to trim the length of your contacts list.

Share Your Contacts with Other Skype Users

You can share selected contacts from your contacts list by sending them to others. This is a nice way to build a community with common interests; for example, a club, family, or work group.

You can use either of two methods to send your contacts list, or a subset of it, to other Skype users:

◆ You can go to **Skype: Contacts: Send Contacts ...**, which displays a window in which you can form a list of your contacts to send to a specified list of Skype recipients (a comma-separated list of Skype user names).

◆ You can selectively choose multiple entries in your contacts list by left-clicking on them while holding down the control key on your keyboard, and then right-clicking on one of the highlighted entries. From the popup menu that appears, choose **Send Contacts ...**, which displays the same window as before, but with the **Send contacts to** field filled in with a comma- (or semicolon-) separated list of the names you selected, as shown in the following figure.

Send contacts to other Skype users.

After you've created a list of contacts, click on the **Send** button. Each recipient of your contacts list will then be asked, via a window that lists the contacts you are sending, whether he or she wants to accept the contacts and add them to his or her contact list. Recipients can choose to accept the whole list or just specific users in the list.

Stop Others from Seeing How Many Contacts You Have

To prevent others from seeing how many contacts you have, go to your online profile (**Skype: File: Edit My Profile ...**), and remove the check mark opposite **Show how many contacts I have.**

Archive Your Contacts

Your contacts list is stored on the Skype network. However, for peace of mind, you might want to create a backup archive of all your contacts to store on your PC. Skype archives your contacts list in *vCard* format, and this is also a good way to export your contacts to another application, many of which can import contacts in vCard format.

def•i•ni•tion

The **vCard** format is a standard by which to exchange personal data, specifically the sorts of data you might find on someone's business card. Skype's archive utility (backup and restore) exports and imports your contacts using the vCard format.

To backup your contacts list, go to **Skype: Tools: Archive Contacts: Backup Contacts to File ...,** and in the window that opens, choose a location and filename under which to store your contacts list and click on the **Save** button. To restore deleted contacts, go to **Skype: Tools: Archive Contacts: Restore Contacts from File ...,** and in the window that opens, navigate to the file in which you stored your contacts list and click on the **Open** button.

Configuring Skype's Options

Skype offers two types of configuration options: those that make something work properly, and those that make something work the way you want. The first is the domain of initial setup, which was already covered earlier in this book, and troubleshooting, for which you should refer to part 5 of this book. The second type is also commonly known as customization, and it will be the focus of this section. That is, in this section, I will only cover those Skype configuration settings that make Skype work how you want it to.

To open the window that is used to configure Skype, go to **Skype: Tools: Options ....** This opens the Skype Options window shown in the following figure. Clicking on one of the category names on the left side of this window displays the configuration settings page for that category. Skype's configuration options are grouped under the following categories: General, Privacy, Notifications, Sounds, Sound Devices, Hotkeys, Connection, Call Forwarding & Voicemail, Video, and Advanced.

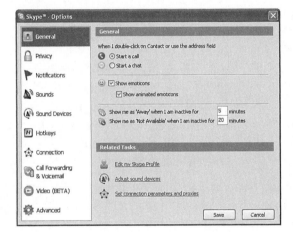

Skype configuration options.

In the following sections I discuss the configuration options in each category that are most relevant to making Skype work the way you want it to. If configuration options for a particular category are dealt with elsewhere in the book (for example, privacy and security is sufficiently important to have its own section, Part 4), I point you to the appropriate section.

General

All the settings under the General category are in one way or another related to making Skype behave the way you want it to. Here are the settings you see when you click on the General category in the Skype Options window:

◆ **When I double-click on Contact or use the address field:**
Selecting **Start A Call** means that whenever you double-click
on an entry in your contacts list, or enter a Skype name in the
address field, you will start a voice call with the contact or named
Skype user. Selecting **Start A Chat** means that the same actions
will start a chat session instead. If you make mostly voice calls,
selecting Start A Call is the better option. Conversely, choosing
Start A Chat is the better option if you chat more often than call.

◆ **Show emoticons** and **Show animated emoticons:** *Emoticons*
are visual icons that convey your emotional state of mind to the
recipient. They are used in chat ses-
sions as shorthand visual replace-
ments for words. You can enable
emoticons for chat sessions by putting a check mark opposite **Show
Emoticons.** Furthermore, if you
have enabled emoticons, you can further enable animated emoticons by
putting a check mark opposite **Show
Animated Emoticons.**

def•i•ni•tion

> **Emoticons** are icons represent-
> ing emotions, such as happi-
> ness, sadness, confusion, and
> the like. Think of them as emo-
> tional icons that communicate
> your mood to the recipient.
> Animated emoticons are simply
> animated versions of static
> emoticons.

◆ **Show me as 'Away'** and **Show me as 'Not Available':** These
settings determine when, after a period of inactivity on your computer, Skype automatically switches your online status to Away or
Not Available. These settings are dealt with in Chapter 15.

◆ **Edit my Skype Profile, Adjust Sound Devices,** and **Set connection parameters and proxies:** These are shortcut links to other
configuration screens.

Privacy

Privacy and security options are dealt with in Part 4 of this book.

Notifications

Notifications are visual cues for events that Skype thinks are important for you to know about, such as when you have an incoming call.
Notifications pop up a window in the lower-right corner of your

desktop for just a few seconds, and then they automatically go away. Clicking on the Notifications category in the Skype options window allows you to enable and disable (that is, put a check mark or remove a check mark from) the following notifications:

♦ **Comes online:** Check this option to get notifications whenever any of your contacts changes his or her status to online. Note that if you have a large contacts list this notification can soon become very tiresome because you will get a lot of notifications, which can be distracting.

♦ **Calls me:** Check this option to receive notification of an incoming call.

♦ **Starts chat with me:** Check this option to receive notification that someone has started a chat session with you.

♦ **Sends me a file:** Check this option to receive notification that someone is sending you a file.

♦ **Requests my contact details:** Check this option to get a notification each time someone requests your contact details.

♦ **Sends me contacts:** Check this option to get a notification when someone sends you a list of contacts.

♦ **Leaves me a voicemail:** Check this option to get a notification of someone leaving you a voicemail message. Note that you don't need to be a Skype Voicemail subscriber to receive voicemails from others.

Also displayed in the notifications screen are:

♦ **Display messages for Help/Tips**: Enables (check) or disables (uncheck) dynamic messages for help and tips.

♦ **Configure sound alerts**: This is a shortcut link to the Sounds configuration options screen (see next section).

Sounds

Skype's sounds are the audible analogs of the visual notifications discussed in the previous section. Sounds can be used instead of, or in addition to, visual notifications. By clicking on the category **Sounds** in the **Skype Options** window, you can choose sounds for several Skype events, including a ringtone for an incoming call, call on hold, incoming chat, and many others. Moreover, you can add your own sounds or purchase additional sounds from Skype (see Chapter 11).

Sound Devices

Setting up your sound devices for Skype is not so much customization, but rather an essential part of running Skype. Also, sound is one of the more problematic areas of Skype; for that reason, tips on troubleshooting Skype sound can be found in Chapter 18.

Briefly, to setup your sound devices for Skype, select your **Audio In**, **Audio Out,** and **Ringing** sound device choices by using the pulldown menus in the **Sound Devices** page of **Skype's configuration options** window. In these pull-down menus you will find the available sound devices for your PC, and a device labeled **Windows default device**. This latter device is equivalent to the voice sound devices currently used by Windows (go to **Start: Control Panel**, double-click on **Sounds and Audio Devices**, and in the window that opens click on the **Voice** tab). However, because the **Windows default device** can change when sound devices are plugged into, and unplugged from, your PC; you would do well to set your sound devices within Skype explicitly. The Sound Devices page of the Skype configuration options window also has check boxes that enable you to have your internal PC speaker (if you have one) ring for incoming calls, and have Skype automatically adjust the settings of your selected Audio In and Audio Out sound devices for you.

Hotkeys

Hotkeys enable you to operate Skype using key strokes (instead of having to use the mouse). This topic is covered in detail in Part 6 of this book.

Connection

Like Sound Devices, options in this category should only be tweaked if you're having problems connecting with the Skype network over your

Internet connection. Troubleshooting Skype networking is dealt with in Chapter 19.

Call Forwarding & Voicemail

Call forwarding and voicemail are dealt with in Chapter 7 of this book.

Video

Configuring and using Skype video are dealt with in Chapter 4 of this book.

Advanced

Clicking on the **Advanced** category in the **Skype Options** window displays more advanced configuration settings. But don't let the term *advanced* scare you off, because there are settings here that all users might find useful when configuring Skype to work the way they want it to. Options under the Advanced category are grouped into the following four subcategories:

◆ **Startup:** Putting a check mark opposite **Start Skype when I start Windows** will ensure Skype is up and running as soon as you logon to Windows. By putting a check mark opposite **Check for updates automatically,** you will be notified when a newer version of the Skype softphone software is available to download.

◆ **Call:** By putting a check mark opposite **Automatically answer incoming calls,** you will in effect pick up all incoming calls without any action on your part. If your work requires your hands to be free at all times, this is a useful feature.

◆ **Chat: Chat style to use** is a pull-down menu that gives you two options: **Skype Default** and **IRC-like Style.** Both options refer to the behavior and look of the chat client for Skype's softphone. The key difference is that IRC-like chat is more terse and message exchanges take up less space than the Skype default style for chat. Putting a check mark against **Show timestamp with chat messages** displays the date and time alongside each message exchanged during a chat session. If you want the chat window to open and display automatically when you receive incoming chat, then put a check mark opposite **Pop up a chat window when someone starts a chat with me.**

♦ **Other:** If you want the window that asks for your permission to share your contact details with others to open and display whenever you get such a request, put a check mark opposite **Automatically pop up requests for my contact details.** By putting a check mark opposite **Enable contact list and history quickfiltering,** when you start to enter a contact name or number into the address bar while in the contacts or history tabs, the list of entries will be filtered and narrowed to show only those that match the text you have so far entered. Some web pages have links, which if clicked, will start a phone call to the person to which the link refers; if you want to be able to click on such links to start a Skype phone call, put a check mark opposite **Associate Skype with callto: links.** Winamp is a popular and free music player for the PC. If you are a Winamp user and want it to automatically pause when making or receiving Skype calls, put a check mark opposite **Automatically pause Winamp during calls.** Finally, by putting a check mark opposite **Display technical call info,** you can hover your mouse over the picture of someone participating in a call (or conference call) to get a popup window with technical details for that call (this information is really only of interest to advanced users).

The Least You Need to Know

♦ You can change the behavior and look of the Skype softphone to be more in keeping with the way you work.

♦ Use your online profile to convey the image and personal details you want made available to the Skype online community, and use your online status to control your visibility among your contacts.

♦ Organize and manage your contacts list to increase your productivity.

♦ Use the Skype Options window to configure its features so that they better meet your needs.

Chapter **11**

Personalizing Skype

In This Chapter

- ◆ Decorating your Skype name
- ◆ Setting your mood message
- ◆ Adding pictures and ringtones
- ◆ Changing your preferred language

The point where configuring Skype to better meet your needs leaves off and where personalization begins is somewhat blurred. Nonetheless, this chapter, in combination with Chapter 10, should cover all of your customization and personalization needs.

You can use Skype's personalization features to change how others in the Skype online community see you, and to alter how you interact with the Skype softphone. So, if you think it's time to add some pizzazz to your online persona, or add some bells and whistles to your softphone, this chapter is for you.

Making Your Skype Name Stand Out from the Crowd

You can decorate your online profile name to make it stand out from other names in a contact list. You can even exert some control over where you appear in those lists: toward the top, or toward the bottom.

To decorate the name others will see for you in their contacts lists (unless, of course, they manually rename your entry in their lists, see Chapter 10), open a word processor such as Microsoft Word. In the word processor, using the keyboard and the symbol box (**Insert: Symbol ...**), construct a name. Remember that, in addition to the letters A through Z and numbers, you can incorporate punctuation marks and nonstandard characters, as shown in the following figure.

Use a word processor to create a name decorated with nonstandard characters.

Once you've created a name, copy it to the clipboard (for example, in Microsoft Word, use the keyboard to type Ctrl+A followed by Ctrl+C; that is, hold down the Control key and press A then C). Next, open your Skype profile by going to **Skype: File: Edit My Profile ...** and paste your decorated name into the field labeled **Full Name** (click on the field and hit the Shift+Ins keys), as shown in the next figure. Then click on the **Update** button.

Paste your decorated name into your Skype profile.

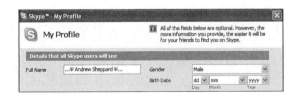

Congratulations, you now have a name that stands out from the crowd in the contact lists of other Skype users, as shown in the following figure.

Note that not all nonstandard characters will work when pasted into the Full Name of your profile, so you'll have to experiment a little. Also, bear in mind that decorated names that begin with any of these characters ! " # $ % & ' ( ) * + ,—. / will appear at the top of a contact list, even before names that begin with a digit. Alternatively, a decorated name that begins with a nonstandard character, such as one of these, Σ Ζ Ψ § £, will appear at or near the bottom of someone's contact list.

How a decorated name appears in another Skype user's contact list.

Set Your Mood Message

A mood message is any short message that people who have you in their contacts list can see. Even though it's called a *mood* message, the message doesn't have to convey your mood.

To set your mood message, click on your name in the notification bar of the Skype softphone and enter a short message in the field opposite your picture, as shown in the following figure.

Set your mood message, or indeed any short message, you want others to see.

Adding Pictures and Ringtones

Adding pictures or *avatars* to your Skype profile gives you a visible online persona (go to **Skype: File: Edit My Profile ...**, and in the window that opens, click on the **Change ...** button). It is the primary means by which you are seen and recognized by others in the Skype online community. Because your Skype picture can be any image, you are free to let your imagination—and ego—run wild!

def•i•ni•tion

> **Avatars** are abstract, graphical depictions of people. They usually emphasize some personal attribute or characteristic. For example, if you want your online persona to reflect your passion for, say, sports, you would choose an avatar that has some sort of sports theme.

Skype also allows you to personalize your interaction with the Skype softphone by associating different sounds with different events (go to **Skype: Tools: Options ...,** and in the window that opens, click on the category **Sounds**). Skype comes with a wide variety of sounds and ringtones but you can also add your own. You can download sounds and ringtones from the Internet or you can create your own.

The Personalise Skype Shop

You can buy pictures and ringtones from the Personalise Skype shop at http://personal.skype.com. Pictures and ringtones cost $1.20 (1 €) apiece. Note that before you can buy pictures or ringtones from the Personalise Skype shop, you must have some Skype credits (Skype credits are also used for SkypeOut, see Chapter 5).

To buy a Skype picture, go to http://personal.skype.com. This displays a page featuring pictures and ringtones. Click on the **Pictures** link that appears on the left side of the page. This displays pictures grouped by category. Click on the name of any picture to see that picture, or click on one of the categories to see all pictures for that category. To purchase a picture displayed in a preview page, click on the **Buy now** link. This takes you to the Skype sign in page, where you must enter your Skype name and password to proceed. Follow the instructions to purchase and download the picture. Once you have successfully downloaded the picture, a My Pictures window will pop up, as shown in the next figure. Double-click on your newly purchased picture and it will be added to your Skype profile.

Something to Try

> Skype offers a type of avatar called a WeeMee, which you can decorate with facial features, hairstyles, clothing, and other characteristics. You can build your own WeeMee for use with Skype by visiting the Personalise Skype shop at personal. skype.com and clicking on the WeeMee link. A personalized WeeMee costs $1.99.

A newly purchased picture, in this case "2.0 Is Here," ready to be added to your Skype profile.

To buy a Skype sound or ringtone, return to http://personal.skype. com. This once again displays the pictures and ringtones page. Click on the **Ringtones** or **Ministry of Sound** links on the left side of the page. This displays some popular ringtones, and also shows categories (such as Classics, Swing, Chill Out and so on) for ringtones. Pick a ringtone, and then click on the listen button to listen to it, or click on a category name to display all the sounds in that category. Once you have found and listened to a sound you want to buy, click on the **Buy now** button. This will take you to the sign in page for Skype; enter your Skype name and password, then click on the **Sign me in** button. On the next web page, put a check mark in the box accepting the terms and conditions of your purchase, then click on the **Confirm purchase** button. Follow the instructions to download, and your new sound will be added to the My Sound Library list in the Sounds page of your Skype Options window, which should open automatically when your download is complete, as shown in the next figure. You are now free to use your new ringtone sound to signal one or more of Skype's many events, including incoming calls.

Skype Options window for sounds.

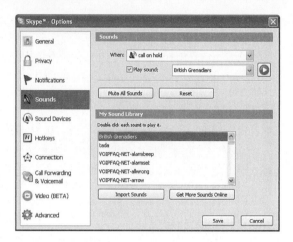

Free Alternatives to the Personalise Skype Shop

Instead of buying pictures and ringtones from the Skype shop, you can download free pictures and ringtones from the web, or you can make your own. The latter option is the ultimate in personalization!

> **Something to Try**
>
> The Windows operating system comes with several free sound files that you can use with Skype. In Skype, go to **Tools: Options ...** and click on the **Sounds** category in the Skype Options window. Next click on the **Import Sounds** button and navigate to **C:\Windows\ Media,** select a .wav sound file, and click the **Open** button. This adds the selected sound to your My Sound Library, making it available for use as an event notification sound or ringtone.

Here are some suggestions for personalizing your Skype softphone:

◆ **Free pictures and ringtones:** Enter "Free Skype (pictures OR sounds OR ringtones)" into any search engine and you are sure to get a lot of useful hits. But to get you started, I'm going to suggest a few websites. For free Skype-compatible pictures and ringtones try: www.themeskype.com, gallery.mobile9.com/c/wav-ringtones/1, and www.voipfaq.net.

◆ **Make your own pictures:** You can make your own picture for use with Skype using any application that can make, or convert, pictures to .png, .jpg, or .bmp formats. Perhaps the simplest way to do this is to use Microsoft Paint, which comes bundled with your Windows operating system (**Start: All Programs: Accessories: Paint**). In Paint, go to **Image: Attributes,** and in the popup window that appears select **Pixels** under Units, set both the width and height of your image to 96, and then click the **OK** button. Use the drawing tools of Paint to create the picture you want. When you're satisfied with the result, go to **Paint: File: Save As …** and in the window that appears, navigate to **My Documents\My Skype Pictures.** From the **Save as type** pull-down menu, select **PNG,** and type a file name for your picture. Click the **Save** button and your picture will be added to those already available for use by Skype. To use the picture you created in your Skype profile, follow the instructions that come later in this chapter for adding a picture to your online profile.

♦ **Make your own sounds:** Sounds and ringtones for Skype are small .wav files. Any application that can generate mono .wav files can be used to create a custom sound or ringtone for use with Skype. Perhaps the easiest way for Windows users to create such sounds is to use the Sound Recorder program that is bundled with Windows (**Start: All Programs: Accessories: Entertainment: Sound Recorder**). Using Sound Recorder you can record sounds you make through your microphone and even mix them with sounds from other .wav files you have on disk (**Sound Recorder: Edit: Mix With File …**). Sound Recorder even comes with some very basic tools for modifying sounds, such as Add Echo and Reverse, available from the Effects menu. Once you're happy with the sound you have made, which should be 30 seconds or less in length, go to **Sound Recorder: Save As …** and in the window that pops up, navigate to **My Documents\My Skype Content.** If the format for the file you are about to save isn't some flavor of mono, click on

> **Caution!**
>
> Skype uses files with a .wav extension, which denotes a Windows audio format. However, the .wav file must meet certain criteria. First, the .wav file must be mono, not stereo. Second, the file size must be less than 1 megabyte, which means that sounds must be short snippets measured in seconds rather than minutes.

the **Change ...** button and make it so. Enter a file name for your sound, and then click on the **Save** button. This will make your new sound available to Skype. To use your newly created sound within the Skype softphone, follow the instructions that appear later in this chapter for associating sounds and ringtones with events within Skype.

Add a Picture to Your Online Profile

To add a picture to your Skype profile or to change an existing picture, you have two alternatives. First, you can display your profile (**Skype: File: Edit My Profile ...**) and then click on the **Change ...** button. Second, you can click on your picture as it appears in the notification bar of the Skype softphone (first, click on the tab with your name on it in the notification bar). In both cases, a My Pictures window will open that displays your library of Skype pictures, as shown in the following figure.

Skype pictures for your online profile.

In the My Pictures window, your current picture is highlighted. To select a different picture, simply double-click on it. It's as simple as that!

> **Caution!**
>
> Pictures that are larger or smaller than 96×96 pixels in size will be scaled appropriately by Skype and may end up looking distorted. To avoid such distortion, you can crop or scale a picture to make it exactly 96×96 pixels in size prior to adding it to your picture library.
>
> Personal pictures work only on a PC where you have added them to your picture library folder. That is, they are not stored on the Skype network, and so if you sign on to Skype from another PC, your profile will use a default Skype picture instead.

Associate Sounds and Ringtones with Events

The Skype softphone has a great many events that might at any moment beg for your attention. In addition to visual notifications (that is, windows popping up), Skype tries to get your attention by using sounds. Moreover, you can associate any sound with any event, so in that way you can distinguish between, say, an incoming call and an incoming chat, and deal with them differently. Skype associates sounds with events using the Sounds window in Skype Options (**Skype: Tools: Options ...** and click on **Sounds**).

You can import sound files on your hard disk for use by Skype by clicking on the **Import Sounds** button.

To associate a specific sound with a specific event, use the **When** pull-down menu to select the event, and use the **Play sound** pull-down menu to select the sound to be played when that event happens. To disable an event notification sound remove the check mark opposite **Play sound**. You can test the sound by pressing the round green button with an arrow on it.

Once you have associated the sounds you want played with certain events, click on the **Save** button. Thereafter, you will receive custom audible notifications for the events of your choice. Note that Skype comes with default sounds for most events, and these can be reset by using the **Reset** button.

You can use the **Mute All Sounds** button to run Skype in silence. (Clicking the **Mute All Sounds** button will change it into an **Enable All Sounds** button, which you can use to enable sounds once again). Finally, clicking on the **Get More Sounds Online** button will take you to the Personalise Skype web shop.

English Not Your Preferred Language? No Problem!

Skype can easily switch between numerous languages. To switch language, go to **Skype: Tools: Change Language,** and from the list that appears, choose your preferred language. All of the Skype softphone menus and buttons will then display text in your preferred language. You can switch back and forth between different languages without restarting Skype.

> ### Something Worth Knowing
>
> At the time of this writing, the following languages were supported by Skype: Arabic, Bulgarian, Chinese Simplified, Chinese Traditional, Czech, Danish, Dutch, English, Estonian, Finnish, French, German, Greek, Hebrew, Hungarian, Italian, Japanese, Korean, Norwegian, Polish, Portuguese, Brazilian Portuguese, Romanian, Russian, Spanish, Swedish, and Turkish.

The Least You Need to Know

- ◆ You can personalize Skype in a variety of different ways.
- ◆ Some ways of personalizing Skype change the way in which you are perceived by the Skype online community.
- ◆ Other ways of personalizing Skype change the way in which you interact with the Skype softphone.
- ◆ You can buy custom pictures and ringtones from the Skype online shop.
- ◆ You can also find free Skype-compatible pictures and ringtones on the Internet. Or make your own!
- ◆ You can easily change the language in which menus and buttons are displayed in the Skype softphone.

12

Skype Toolbars and Buttons

In This Chapter

◆ Making sense of Skype toolbars and buttons

◆ Using the Skype toolbar for Microsoft Outlook

◆ Using the Skype toolbar for Microsoft Internet Explorer

◆ Using Skype buttons

Skype toolbars are productivity tools that integrate with the Microsoft Outlook e-mail application and the Microsoft Internet Explorer web browser. They simplify many communication tasks that involve Skype, all without the hassle of leaving Outlook or Internet Explorer.

Skype buttons are also productivity tools, in that they make it easier for you to reach out to others using Skype, and at the same time they make it easier for others to contact you. Think of Skype buttons as the online equivalent of a business card. That is, they make it easier for people to remember and contact you.

In short, Skype toolbars and buttons aim to reduce the "friction" inherent in using a tool such as Skype. This chapter shows you how to raise your productivity a notch or two by using these free tools.

The Skype Toolbar for Microsoft Outlook

The Skype toolbar adds some extra graphical interface items to Outlook, as shown in the following figure, and also adds some new functionality. Using the Skype toolbar you can integrate the call and chat functions of Skype with the e-mail, contacts, and time-management functions of Microsoft Outlook.

The Skype toolbar for Microsoft Outlook adds new graphical interface items and new functionality.

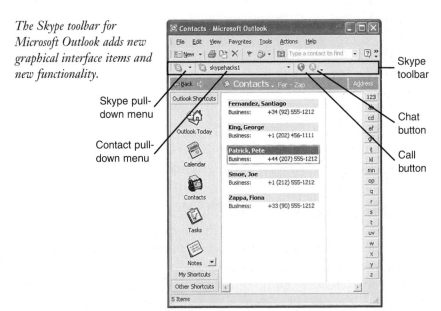

Skype toolbar

Skype pull-down menu

Contact pull-down menu

Chat button

Call button

Download, Install, and Configure the Toolbar

To download and install the Skype toolbar for Outlook, first make sure that Outlook is not running. Next, go to **www.skype.com** and click on **Download** on the menu at the top of the page. Click on the **Skype Email Toolbar** link, which will take you to the Skype Email Toolbar page. Next, click on the **Get it now** button, and then follow the detailed instructions on the Downloading Skype Toolbar for Outlook page. When you are finished, the toolbar will be installed on your machine.

When you start Outlook after downloading the Skype Email Toolbar, a window will pop up asking you to give the toolbar access to Skype, as shown in the following figure. This is a security feature to prevent malicious tools from making use of Skype. For the toolbar to work, you will have to select **Allow this program to use Skype** and click on the **OK** button.

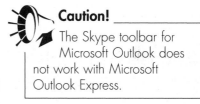

Caution!

The Skype toolbar for Microsoft Outlook does not work with Microsoft Outlook Express.

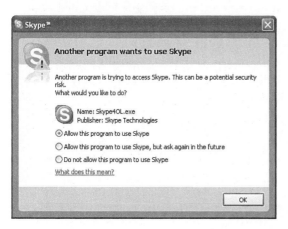

You must give your permission for the toolbar to access Skype.

Next, the Skype for Outlook Configure window will open, as shown in the next figure. You can also open this window at any time by going to **Outlook**, clicking on the **Skype** pull-down menu (denoted by a blue icon with an "S" on it), and choosing **Configure ....**

The Skype toolbar for Microsoft Outlook configuration window.

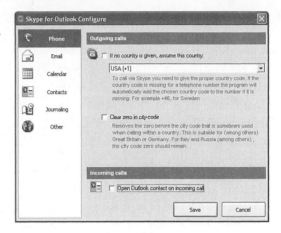

You can use the Skype for Outlook Configure window to configure the behavior of the toolbar in the following ways:

♦ **Phone:** For outgoing calls you can set the assumed country code for telephone numbers in your Outlook contact list that have no explicit country code (so that a number such as 203-555-1212 will be dialed correctly in Skype as +1-203-555-1212), and you can also specify that the leading zero for a city code is dropped (so, for example, the London, U.K., number 0207-555-1212 will be correctly dialed as +44-207-555-1212, provided you have set up an assumed country code of 44). For incoming calls you can set things up so that if the call is from a contact you already have in Outlook, that contact's details will automatically open and be displayed by Outlook.

♦ **E-mail:** Put a check mark against **Show Toolbar For E-mail,** if you want the Skype toolbar to be displayed when you read your e-mail. Provided you have enabled the Skype toolbar to be displayed when reading Outlook e-mail, you can also choose to have the toolbar analyze incoming e-mails for telephone numbers and links to Skype; you can also choose to set things up so that if you start a chat session with the sender of an e-mail, the title for that chat session would be the same as the subject line of the e-mail.

♦ **Calendar:** Put a check mark against **Calendar** if you want the Skype toolbar to be displayed when in the calendar function of Microsoft Outlook. You can also choose to change your Skype online status (to Away, Invisible, Not Available, or Don't Disturb) whenever your Outlook calendar has you in a meeting.

♦ **Contacts:** By putting a check mark against **Automatically Update Outlook Contacts With Skype Names,** the Skype toolbar will search your Outlook contacts and match them against your Skype contacts. Whenever an exact match is found, your Outlook contacts will be updated automatically with their corresponding Skype details. Putting a check mark against **Search E-mail Addresses in Skype Directory,** the Skype online directory of users will be searched using the e-mail addresses found in your Outlook contact list. Whenever a direct match is found, your Outlook contact list will be updated automatically with the Skype user details. Finally, by putting a check mark against **Create New Outlook Contacts,** whenever a Skype user is found that does not yet have an entry in your Outlook contact list, a new Outlook contact for the Skype user will be created.

♦ **Journaling:** By putting a check mark against **When a New Skype Call is Made Create Journal Item Log,** whenever you make a call using Skype, a journal entry for that call will be made in Outlook. Similarly, by putting a check mark against **Create Outlook Journal Item on Incoming Call,** then an Outlook journal entry will be made for every incoming Skype call. Using this window form, you can also specify where the journal folder should be stored.

♦ **Language:** Using a pull-down menu, you can choose the display language of the Skype toolbar: Chinese Simplified, Chinese Traditional, Dutch, English, Estonian, French, German, Italian, Japanese, Korean, Polish, Portuguese, Russian, Spanish, and Swedish. Note that you can set the display language for the toolbar independently from the language for the Skype softphone.

Using the Toolbar

The Skype toolbar is displayed alongside Outlook's own toolbars. It also adds some items to the Outlook menu and adds some options to the right-click menu.

The Skype toolbar consists of three elements: a Skype pull-down menu, a contact pull-down menu, and call and chat buttons. These elements are shown in the following figure.

Skype toolbar elements (left to right): Skype pull-down menu, contact pull-down menu, call and chat buttons.

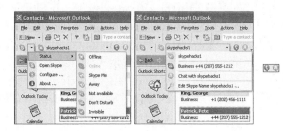

Let's look at what you can do with each element:

◆ **Skype pull-down menu:** Using this menu, you can change your online status for Skype, open Skype (so that the Skype softphone window pops up), open the Skype toolbar configuration window, and display an About window that tells you what version of the toolbar you are running.

◆ **Contact pull-down menu:** If your Outlook contact is also a Skype user, there will typically be four menu items. First, there's the contact's Skype name and online status; clicking on this menu item will start a Skype call to the named user. The second menu items list any telephone numbers that belong to the contact; clicking on one of these entries will start a SkypeOut call to the chosen number. Third, clicking on **Chat with ...** the named Skype user will start a chat session. Fourth, clicking on the **Edit Skype Name** link opens a window that allows you to change the Skype user name associated with an Outlook contact.

When one of your Outlook contacts is not a Skype user, the options available through the contact pull-down menu change. The first menu items list any telephone numbers that belong to the named contact. When you click on the second menu item, **Find Skype Name,** a window opens that enables you to search for that contact in the Skype online directory, or otherwise associate that contact with a Skype name.

◆ **Call and chat buttons:** Whenever these buttons are active (not grayed out), you can click on them to start a call or chat with the Skype user displayed in the contact pull-down menu.

Note that the Skype toolbar for Microsoft Outlook also adds an item to the main menu of Outlook. By going to **Actions: Skype** in Outlook, you will be presented with the same menu options as are available from the toolbar contact pull-down menu. The same is true if you right-click on a contact and select **Skype For Outlook** from the popup menu.

Functions available via the Skype toolbar for Microsoft Outlook depend upon where in Outlook you happen to be:

- ◆ **Outlook Today:** Only the Skype pull-down menu is available. The other elements of the Skype toolbar are inactive and grayed out.

- ◆ **Calendar:** Provided you have enabled the display of the Skype toolbar within the calendar function of Outlook, you will see the Skype pull-down menu. However, the contact pull-down menu and call and chat buttons are inactive and grayed out.

- ◆ **Contacts:** The whole Skype toolbar is active.

- ◆ **Tasks:** The Skype toolbar is not displayed.

- ◆ **Notes:** The Skype toolbar is not displayed.

- ◆ **E-mail Inbox and Opened E-mail:** Provided you have enabled the display of the Skype toolbar for e-mail and also enabled all other toolbar options for e-mail, you will see something like the following figure. By enabling **Analyze Message Text,** the message will be parsed for telephone numbers and links to Skype user names, and these will appear in the contacts pull-down menu, from which you can call and chat with people whose contact details are embedded within the e-mail message. By using the Skype toolbar for Outlook, replying to an e-mail by voice or chat is just a click away!

Something to Try

If you want to include a link to your Skype name in an e-mail message, enter the following HTML into the body of the message: <a href="skype:tcigtskype"> skype:tcigtskype </a> (replacing the tcigtskype with your own user name). Alternatively, when sending an e-mail using Microsoft Outlook, go to **Insert: Hyperlink ...** and in the address box of the window that pops up, enter: skype:tcigtskype (again, replacing tcigtskype with your user name).

Skype toolbar for an open e-mail message.

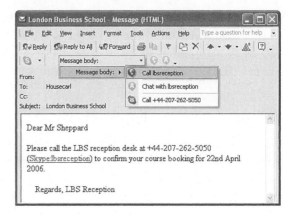

The Skype Toolbar for Microsoft Internet Explorer

The Skype toolbar for Internet Explorer adds a toolbar (but no main menu items), as shown in the following figure. Using the toolbar, you can more effectively use Skype as a communication tool while surfing the web.

The Skype toolbar for Microsoft Internet Explorer adds new menu items and new functionality.

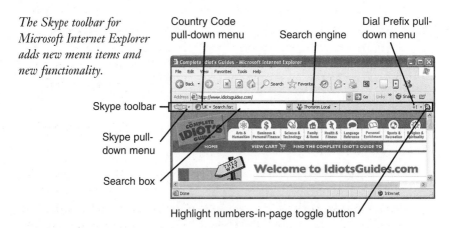

Download, Install, and Configure the Toolbar

To download and install the Skype toolbar for Internet Explorer, go to www.skype.com and click on **Products** on the menu at the top of the page. Next, click on the **Skype Web Toolbar** link, which will take you to the Skype Web Toolbar page. Click on the **Get it now** button, and then follow the detailed instructions for downloading the Skype toolbar for Internet Explorer. This will get you to the stage where the toolbar is installed on your machine.

At the end of the installation, Internet Explorer will be reopened and you will be presented with a short tutorial on how to use the Skype toolbar.

To configure the Skype toolbar for Internet Explorer (IE), go to **IE**, click on the **Skype** pull-down menu (denoted by a blue icon with Skype on it) and then click on **Options ...**, which will enable you to do the following tasks: Check or uncheck Autocomplete Search Fields, Clear Search Fields (that is, clear the search box), Check for Updates to the toolbar, and Uninstall the toolbar.

In comparison with the Skype toolbar for Outlook, the configuration options for the Skype toolbar for Internet Explorer are rather limited.

Using the Toolbar

Before describing how to use the Skype toolbar for Internet Explorer, I must first tell you that three elements of the Skype toolbar have nothing whatsoever to do with Skype! The three elements are the Country Code pull-down menu, search box, and the search engine. These are simply a means by which Skype can get paid for searches you carry out through the sponsoring search engines. You might well find them useful, but they don't have anything to do with using Skype and so will not be discussed in this book.

Here are the three remaining, and useful, elements of the Skype toolbar for Microsoft Internet Explorer:

♦ **Skype pull-down menu:** Using this menu you can open the Skype softphone (so that its main window pops up), change the configuration options for the toolbar, get access to help by displaying the Skype toolbar tutorial in the web browser, and display an About window that tells you what version of the toolbar you are running.

♦ **Dial prefix pull-down menu:** You can use this menu to set the default country code dialing prefix for telephone numbers in a web page that are missing their country code prefix.

♦ **Highlight numbers-in-page toggle button:** Use this button to toggle on and off the Skype toolbar feature that parses web pages to highlight numbers it thinks are telephone numbers. When your mouse hovers over a highlighted number, a popup appears which, when clicked, will start a SkypeOut call to the number in question, as shown in the next figure.

Highlighted numbers in a web page can be used to start a SkypeOut call.

Skype Buttons

When you add Skype buttons to your web pages, blogs, or your e-mail signature, other Skype users can click on your button to call you. Skype buttons are a good way to say: hey, call me!

Caution!

If you use a Skype button that shows your online status as part of your e-mail signature, then whether that button accurately reflects your online status will depend on the e-mail client of the recipient. Some e-mail clients update the message body when an e-mail is opened, others do not. For this reason I suggest you use a Skype button that doesn't show online status for your e-mail signature.

If you want your Skype button to accurately reflect your online status, you will also have to change your Skype privacy setting. Go to **Skype: Tools: Options** … and in the window that opens, click on the **Privacy** category. Then put a check mark opposite **Allow my status to be shown on the web**, and click the save button.

There are two kinds of Skype buttons, as shown in the following figure: those that show your online status, and those that do not. In the case of the former, simply by viewing your button people will know whether you're currently online. Buttons also come in different sizes, styles, and colors.

Skype buttons. The button on the left does not show online status, while the button on the right does.

You can make buttons that invite the person who sees them to do other things. Here are the options: Call me!, Add me to Skype (that is, ask to be added to someone's contacts list), Chat with me, View my profile, Leave me voicemail, or Send me a file.

To get your own Skype button, go to www.skype.com/share/buttons. In the Web page that appears, enter your Skype name and select the type of button you want. A preview of your button should appear in the web page, and the text box below Copy & Paste This Code should be updated to contain the *HTML* (web or e-mail, you must choose) to create the button. Right-click on the text box and select **Copy** from the popup menu. You can now paste the HTML directly into the HTML of a web or blog page, or into an e-mail signature. I describe how to do this in the following sections.

def•i•ni•tion

HTML stands for hypertext markup language. This is a fairly simple text-based language that describes to web browsers how a web page should be displayed. In HTML, text elements are given properties by enclosing them between tags; for example, <i>Italic Text</i> makes the text between the two tags <i> and </i> appear as italic.

Add a Skype Button to Your Web or Blog Page

To put a Skype button on your web page, open the page in your favorite HTML editor and paste the HTML generated for the button (see above) into the page at the point where you would like it to be displayed. Similarly, most blog editors allow you to insert raw HTML into your blog page, so switch to HTML mode and insert the Skype button HTML text where you want it to appear. It's as simple as that!

Add a Skype Button to Your E-mail

What follows is the procedure for Microsoft Outlook. However, most modern e-mail clients normally allow you to create an e-mail signature that incorporates HTML.

Open Notepad (**Start: All Programs: Accessories: Notepad**), paste the HTML for your button into it, and then save it as a file somewhere convenient (for example, in the folder My Documents) as filename.htm; but replace "filename" with a name of your choice.

Now, in Outlook, go to **Tools: Options ...** and in the window that appears, click on the **Mail Format** tab. Under the Signature category, click on the **Signatures ...** button. This displays the **Create Signature Window.** Click on the **New ...** button. Enter a name for your signature, and click on **Use This File As A Template** and browse to filename. htm. Click the **Next** button. Add whatever additional text you want to your signature and click on the **Finish** button. Keep clicking **OK** until you're back to Outlook.

Something to Try

The web page www. skype.com/share/buttons/ wizard.html allows you to construct Skype buttons. This page gives you more control over how the buttons look and work.

From now on, any e-mail you send to others will include your Skype button in the signature.

The Least You Need to Know

◆ Skype toolbars and Skype buttons are productivity tools. Even better, they're free!

◆ The Skype toolbar for Microsoft Outlook integrates Skype into Outlook's contacts and e-mail functions.

◆ With the Skype toolbar for Microsoft Internet Explorer, you can use Skype directly from web pages while surfing the web.

◆ You can embed Skype buttons in your web pages, your blog, or your e-mail signature, so that people can more easily contact you.

Chapter 13

Skype Add-Ons

In This Chapter

♦ Understanding what Skype add-ons can do for you

♦ Finding Skype add-ons

♦ Using hardware add-ons for Skype

♦ Using software add-ons for Skype

Skype comes with many useful functions, factory installed, so to speak. Just by downloading and installing Skype on your PC, you get a wealth of useful, and free, functionality.

However, like any application, Skype cannot be all things to all people. For that reason, Skype wisely enabled its softphone to be enhanced by means of *add-ons*. By using software or hardware add-ons for Skype, you can significantly increase the number of things Skype can do. Some add-ons you have to pay for, but there are also a surprising number of free software add-ons available. As a matter of fact, you've already learned about two free software add-ons: the toolbars for Microsoft Outlook and Microsoft Internet Explorer described in Chapter 12.

def•i•ni•tion

Add-ons are software applications or computer hardware that run independently of Skype but that enhance what you can do with Skype. Also known as plug-ins or add-ins (sadly, the terminology isn't very concrete or precise), they typically use Skype's Application Programmer's Interface, or API.

In this chapter you will learn about the types of Skype add-ons that are currently available and where you can find them. As you read about these features, keep in mind that new add-ons are being made available all the time, and this chapter cannot possibly cover them all. However, after reading this chapter you should be able to find add-ons that meet your particular needs. Don't be afraid to experiment!

Note that managing the security aspects of Skype add-ons is dealt with in Chapter 16.

Where to Find Skype Add-Ons

The first place you should start looking for Skype add-ons is in the extras gallery of Skype's website, which can be found at share.skype.com/directory. Here you will find add-ons grouped by category and rated by people who actually use the add-ons. This star ranking system is a good indicator of the usefulness and quality of each add-on.

Skype user forums are also good places to find out about add-ons. Specifically, I suggest you try the Skype extras forum at forum.skype.com/viewforum.php?f=27. Here you will find information about Skype hardware and software add-ons. However, be aware that much of the information is posted by the vendors themselves, and so it is somewhat biased. If you are of a mind to use a particular add-on, a more general search of all the forums using the name of the add-on will often turn up a wealth of opinions posted by members of the Skype user community.

A less targeted, but nonetheless worthwhile, approach is to run a Google search on the term: Skype "add-on" OR addon OR "plug-in" OR plugin OR "add-in" OR addin. This will result in a lot of hits; but as they're ranked by relevance, the first few pages of results should prove useful in your search for Skype add-ons.

Finally, if you've been unable to find what you want using the methods suggested so far, feel free to post your request to the Skype user forums. You can typically expect to get an intelligent response within 24 hours, regardless of how esoteric your question—try that with your regular phone company!

Something Worth Knowing
You don't need to register to browse the Skype user forums. However, if you want to post questions, you will have to register with the forum. Registering should take only a minute. Moreover, the only information you are required to provide is your user name (this can be different from your Skype user name), a valid e-mail address, and a password. That's it. If you want to give more details, you can, but it's not required.

A Quick Tour of Hardware Add-Ons for Skype

A Skype hardware add-on is an item of computer hardware that is specifically targeted for use with Skype. So, for example, a USB headset isn't a Skype add-on in the sense that it can be used as a general audio device for your PC, whereas a Skype-to-phone adapter that enables you to use a regular phone handset with Skype is a Skype add-on.

As Skype grows in popularity, more and more Skype hardware add-ons are coming to market. Already, there is a bewildering array of innovative hardware that extends Skype in one way or another. The following figures show, from left to right: a WiFi Skype phone that can be used as a wireless handset in your home or as a mobile phone while on the move (provided you can connect to a wireless network), a Skype-to-phone adapter that plugs into a USB port on your PC and into which you can plug a regular phone, and a mouse that also functions as a phone handset.

The market for Skype hardware add-ons is still quite young, so I leave it to your imagination as to what the future holds. A Skype-enabled cappuccino-making machine perhaps? Take my word for it, Skype hardware add-ons are going to be big!

Some examples of Skype hardware add-ons: a WiFi handset, a Skype-to-phone adapter, and a mouse that is also a handset!

A Quick Tour of Software Add-Ons for Skype

A Skype software add-on is an application that you install and run on your PC alongside Skype. You can use add-ons to control Skype and, likewise, Skype can control the application. It's a very flexible arrangement that enables an add-on to do pretty much anything you can imagine you might want to do with a phone call, chat session, or video call.

In this section I take you on a whirlwind tour of the add-ons I think are most useful. This overview will give you a sense of the range of add-ons currently available, and perhaps whet your appetite enough to try some, and maybe even search for more.

Here is a list of some of the most useful (well, interesting, at least) software add-ons for Skype:

♦ **Pamela** (www.pamela-systems.com; around $20): This veritable Swiss army knife of an add-on includes features such as call recording, forwarding voicemail via e-mail, podcasting support, automatic answering of calls and chat, and automatic changing of Skype online status when in a call.

♦ **Festoon** (www.festooninc.com; free): Skype is presently limited to one-on-one video calls. The Festoon add-on expands video calls to video conferencing, so you and many others can talk face-to-face at the same time. Festoon also enables you to share your desktop with others.

♦ **SkyLook** (www.skylook.biz; free for the first 14 days, then $34.95 for a home user license): An add-on for Microsoft Outlook that has quite a few more bells and whistles than Skype's own free toolbar for Outlook (see Chapter 12). SkyLook includes a replacement for Skype voicemail, call recording and archiving, and many other features.

◆ **KishKish** (www.kishkish.com; some versions are free, others must be purchased): KishKish SMS is a free service that enables GSM mobile phone users to send chat messages to you on Skype. KishKish Mobile is a mobile phone gateway that enables you to call your Skype contacts from your mobile phone. KishKish SAM is a voice answering machine with video greetings and other neat features. KishKish Book is a utility to help you organize your Skype contacts (as an alternative to the contacts grouping feature of the Skype softphone; see Chapter 10).

◆ **EQO** (www.eqo.com; SkypeOut and mobile carrier charges apply, but EQO is currently free): EQO in effect extends Skype to your mobile phone. It brings your Skype contacts to your mobile phone and can route incoming and outgoing calls via Skype to and from your mobile phone. EQO is clearly targeted at people on the move who want to stay in touch with their Skype buddies.

◆ **Jyve** (www.jyve.com; free, but requires registration): Jyve is both a set of tools and a Skype-centric online community. The Jyve community is where you can go to find other Skype users with similar interests to you, as the community is centered around interest groups; for example, if you're interested in learning and practicing a foreign language with native speakers, you will find language interest groups for this purpose on Jyve.

◆ **Talking Headz** (www.gizmoz.com; must be purchased): This add-on belongs in the "for fun" category. Its animated avatars speak the words exchanged during a chat session and can perform other interesting antics as well.

The Least You Need to Know

◆ Skype's functionality can be extended by using hardware and software add-ons from third-party vendors.

◆ Many of the software add-ons for Skype are free.

◆ Add-ons for Skype range from serious productivity tools to fun extras.

◆ Finding add-ons that match your interests and needs is simply a case of knowing where to look on the Internet.

◆ As Skype has grown in popularity, the number and types of add-ons has also grown.

Part 4

Privacy and Security

Privacy and security are everyday concerns for all PC users. Fortunately, Skype privacy and security issues are easy to understand and, more importantly, easy to address through common-sense policies and configuration settings for the Skype softphone. Follow the guidelines in this section and you can use Skype with the levels of privacy and security that are appropriate to your needs.

Chapter 14

Privacy and Security Pitfalls

In This Chapter

♦ Avoiding privacy and security pitfalls

♦ Identifying threats to your privacy

♦ Identifying threats to your security

♦ Setting your privacy and security profiles

♦ Letting your kids use Skype

Privacy and security are complex subjects, but important ones nonetheless. Moreover, they're areas in which people don't always think and behave rationally. My goal in this chapter, as well as in Chapters 15 and 16, is to provide you with some concrete steps so you can feel safe and secure while using Skype.

First, though, let's stop and consider what the terms *privacy* and *security* mean in the context of using Skype. When I refer to privacy, I'm talking about control over your visibility within the Skype online community, and control over the amount of personal information you share with others. Security, on the other hand, means freedom from anxiety or fear that your PC will come to harm by using Skype.

And as one final thought before moving forward, keep in mind that a PC running Microsoft Windows can hardly be considered a secure platform! It seems that every day some new security advisory is posted on the Internet that highlights a privacy or security flaw in Windows or one of Microsoft's applications, such as Internet Explorer or Outlook. So, when deciding on your desired level of privacy and security for Skype, keep things in perspective, because many of the threats you will face come not from Skype, but from elsewhere.

> **Something Worth Knowing**
>
> You can check for Skype security bulletins and obtain more information on what Skype is doing to ensure your privacy and security by visiting the Skype Security Resource Center at www.skype.com/security.

Privacy and Security as Trade-Offs

Let me be blunt. There are no such things as total privacy, or total security. Each is a trade-off in terms of the effort you want to expend to achieve them versus the benefits of using a technology like Skype. There is no doubt that by using Skype you put your privacy and security at some degree of risk, but to forgo the use of Skype is to give up a lot of benefits as well. So, in the final analysis, it's a trade-off between the benefits and the risks.

That said, you can nevertheless make intelligent choices about how you use Skype that strikes the right trade-off—that is, a trade-off you can comfortably live with. It is also the case that some configurations of Skype are more privacy-friendly and security-friendly than others. Indeed, by putting a few simple check marks in the appropriate boxes of Skype's configuration options window, you might achieve most— perhaps all—of your privacy and security goals.

Threats to Privacy

Privacy threats are evolving all the time, and so I cannot comprehensively cover every type of threat to your privacy. However, I can point out potential sources of threats. This is useful both from the point of view of being able to judge for yourself the severity of the threat, and knowing from what quarters such threats are likely to come.

My goal is to provide you with a framework within which to make informed decisions about privacy when using Skype. And again I remind you that there isn't any such thing as total privacy, so making reasonable and informed decisions is the best you can do.

Here is a summary of some of the threats to your privacy you might face when using Skype:

♦ Voice *spam* (also known as spam over internet telephony, or SPIT) are unsolicited and unwanted calls from others. Think, phone calls from double-glazing salespeople, or worse!

> **def•i•ni•tion**
>
> **Spam** is the term used to describe any message sent via voice, chat, or e-mail that is both unsolicited, and unwanted by the recipient. Voice spam is also known as spam over internet telephony, or simply SPIT.

♦ Your online status can let others know when you are sitting at your PC and, equally important, when you are not! Employers might be upset to find, for example, that you're not online for much of the workday.

♦ Your online profile contains fields into which you can enter personal information, some of which will be made available to other Skype users. You need to think long and hard about what to put into your online profile because this is what the Skype community will see.

♦ Skype enables video by default. So if your PC has a webcam, when you pickup a call, whatever your webcam sees is broadcast to the caller.

♦ Hijacking of your account. Although this is very difficult from a technical point of view, from a human-exploit point of view it is all too easy. If you spend long periods away from your computer while leaving Skype running and your account signed in, pretty much any passer-by can use your account. This problem is made worse by the fact that you can configure Skype to automatically start and sign in to your account when your PC is powered on.

◆ Impersonating you as a Skype user is again technically very difficult to do, but simple tricks can nevertheless be quite effective. For example, if someone registers a Skype user name the same as yours, but merely with a "." (period) at the end, it can easily be mistaken by others for the genuine you. Impersonating another Skype user for the purposes of deceiving someone is illustrated in the next figure (note the use of the trailing period ".").

◆ A carelessly chosen Skype user name can give away your gender, age, nationality, ethnicity, or location. A Skype user name with "babe" in it is almost certainly going to get you a lot of unwanted attention.

◆ *Spyware*, *adware*, and *malware* are all items of software that unscrupulous vendors bundle with another desired piece of software. Skype has made a commitment to supply its Skype softphone free of all spyware, adware, and malware. But Skype is a commercial company that faces commercial pressures, so it's always a good idea to check the end user license agreement (www.skype.com/company/legal/eula/index.html) and privacy policy (www.skype.com/company/legal/privacy/privacy_general.html) from time to time. I certainly hope that Skype sticks to its excellent policy. But how does that saying go … ah, yes, "trust, but verify."

def•i•ni•tion

Spyware is software that—unknown to you—monitors what you do on your PC and reports such activity to someone else. **Adware** irritates you from time to time with—usually unwanted—advertisements for products and services. **Malware** is any software that damages or degrades the performance of your PC, or steals or damages your data in any way.

◆ Censorship is an ever-present threat to everyone's privacy and freedom of expression. Skype has put word filters into its softphone chat function at the behest of the Chinese government (that is, certain words that the Chinese government finds offensive are automatically removed by Skype). The details of this policy are somewhat scanty, but it's clearly a worrying sign.

A request from another Skype user for you to share your contact details. In this case, it's an attempt at impersonation!

Threats to Security

Security threats, like those for privacy, are continuously evolving. So, again, the threats I describe in this section are merely illustrative of the things you should pay attention to when making decisions about Skype security.

Here are some security threats you should know about:

+ Skype allows you to receive files from other Skype users with whom you have shared your details. Even though people with whom you share details are typically known to you—or, rather you should know all the people with whom you share details— there is no guarantee that their computers and the files they send to you aren't infected with a virus. Whenever you download a file from the Internet, or receive a file via e-mail or Skype file transfer, there's always the possibility that it contains a virus.

+ Skype may conflict with other applications running on your PC, which may render Skype, the other application, or your PC unusable. Not a common occurrence, but something to watch out for.

+ Skype uses the Internet to make and receive calls. So, if your Internet connection stops working, then you can't use Skype. Even though Internet services are reliable nowadays, it will be some time—possibly never—before they can rival the reliability of the regular telephone network, which can have on the order of 99.9 percent availability (about five minutes of downtime per year). If you keep a regular telephone for 911 emergency services in addition to Skype, this is not a problem as you can always fall back to the regular phone. This isn't a security risk of Skype per se, but a failure of your Internet connection will nevertheless stop Skype dead.

◆ Skype is a fairly new product that is evolving rapidly. Already there have been versions of the Skype softphone with security vulnerabilities which, thankfully, were patched very quickly by Skype. Skype's P2P network is likewise new and, even though it so far appears not to have been hacked, it will clearly become all the more attractive a target for hackers as Skype grows in popularity. Skype uses end-to-end encryption to secure its voice, video, chat message, and file transfer traffic from your PC to the recipient's PC. So for Skype-to-Skype calls and chat, your conversation or message exchange is more secure than when using the public telephone network. However, when you make a SkypeOut call to a regular or mobile phone, your call is typically not encrypted from the time it leaves the Skype network until it reaches the destination phone (though some mobile networks do encrypt calls).

Privacy and Security Profiles

To simplify the task of deciding on your privacy and security goals, and the actions you need to take to achieve them, I have created the following profiles:

◆ **Low Security & Privacy Profile:** You are rather indifferent to this whole privacy and security thing! Your goal is to use Skype to talk with friends and family, and perhaps to also make new friends in the Skype online community. You don't have particularly strident political views, and what you talk about on Skype is no different than what you talk about in a café or bar. You feel that there's nothing on your PC worth stealing and, if worse comes to the worst, you'll just wipe the hard disk and reinstall Windows and everything else.

♦ **Medium Security & Privacy Profile:** You want some protections and safeguards in place for your PC to avoid theft of sensitive documents and to maintain the integrity of your computer (the thought of reinstalling Windows just fills you with dread). You're also somewhat sensitive to privacy issues in that you like to keep some conversations private and want to be selective about what others know about you.

♦ **High Security & Privacy Profile:** Business people demand higher levels of privacy and security, and so it is with you. Perhaps you work in human resources, so privacy is key to your day-to-day functioning. Or perhaps you work in a patent and intellectual-property department, so security is essential. You have a reasonable level of paranoia about privacy and security because your job depends on it!

In the following chapters I will use these profiles to help you configure Skype to achieve the level of privacy and security appropriate to your needs and wishes.

After determining which profile best describes you, you can take the appropriate actions (see Chapters 15 and 16) to achieve the required level of privacy and security.

Skype and Kids

Although Skype doesn't offer any built-in parental controls, there are some common sense things you can do to help improve the privacy and security of your children if they use Skype.

The first thing you should do is find out if your kids are already using Skype. If they are using Skype, find out who they are talking to and set some guidelines for who they can talk to, when, and how often.

Checking to See if Skype Is Installed in a Computer

Here's how to find out if Skype is installed on a computer.

Go to **Start: Control Panel** and in the window that opens, double-click on **Add or Remove Programs**. This opens a window listing all the programs that are installed on that PC. Scroll down the list, looking for Skype. If such an entry exists, Skype is installed. If you want to uninstall Skype, highlight the entry by clicking on it, then press the **Remove** button and follow any instructions given.

Looking Over the Contact List

Check to see who is in your kids' contact lists. Names and mood messages offer a lot of clues about the kinds of people your son or daughter are communicating with. References to sex or violence in someone's name or mood message should at least raise an eyebrow, if not set off alarm bells!

Don't hesitate to ask your kids about their contacts. And, if you're really curious, call or chat with the contact yourself. Finally, remember that you can block undesirable contacts (see Chapter 15).

Reviewing Call and Chat Histories

Events in the Skype history window can be deleted individually, or the whole history can be cleared. Likewise, past chat sessions can be retained or discarded. Moreover, anyone can setup and use any number of Skype usernames, and switching from one to another is as simple as signing off as one, and signing in as another. Nevertheless, looking over your kids' call and chat history might give you some idea of who they are talking to, and in the case of chat, about what.

The Least You Need to Know

- Skype has worked hard to provide its users with reasonable levels of privacy and security.

- There's no such thing as total privacy or total security, and so you have to strike a balance between the features of Skype you find useful and your desired level of privacy and security.

- By choosing one of the privacy and security profiles described in this chapter, you will simplify the actions you need to take in chapters 15 and 16 to get the level of each you desire.

- Skype is popular with kids, and there's a lot you can do to protect their privacy and security while still letting them use Skype.

Chapter 15

Managing Your Privacy

In This Chapter

◆ Choosing and using a username and password

◆ Reviewing Skype's privacy settings

◆ Adjusting your privacy settings to match your privacy profile

The goal of this chapter is to show you how to change those settings in Skype that most affect your privacy. After a general overview of Skype's privacy settings, I explain what settings to use to achieve three different levels of privacy (from low to high) based on the three privacy profiles described in Chapter 14.

Even if you want to customize your own privacy settings, starting with one of the three privacy profile templates will give you a sensible starting point from which you can then twiddle and tweak settings until you're happy. Getting the privacy settings for Skype set to the right level for your own sense of privacy is important because they can, in large part, determine your enjoyment of Skype. With the right privacy settings in place, you can simply enjoy using Skype and let privacy concerns ride in the back seat!

Choosing a User Name and Password

You can begin to eliminate privacy problems at the very outset by choosing an appropriate user name.

Before moving on to a checklist of what makes for a good Skype user name, let's first review the rules for constructing a valid Skype name. Such a name …

- Must be at least 6 characters long and, at most, 32 characters.

- Must start with a letter (digits—0 through 9—can be used anywhere, except as the first character).

- Cannot have any spaces or tabs.

- Can include underscores (_), hyphens (-), periods (.), and commas (,).

- Cannot include the following characters: ! @ # $ % ^ & * ( ) [ ] { } + = | \ ~ ` : ; " ' < > ? /

> **Caution!**
>
> Here's a cautionary tale about choosing an appropriate Skype user name. When I set up a Skype user account for my wife, I chose for her a user name that was her nickname at work. That name happened to include the word "girl" in it. Within four hours of setting up the name, she had received five authorization requests, all clearly from men, all clearly requesting some form of intimate communication. This is not to say that Skype has a disproportionate number of such people, only that all online communities—and Skype's online community has grown very large—have a small minority of such reprehensible people. The good news is that Skype enables you to very easily block these users, but if you get large numbers of authorization requests, this can become tedious.

Use the following checklist to choose a Skype user name, and you'll have one fewer privacy concern to worry about. Avoid choosing a user name that tells people …

- What sex you are: Avoid jane_mary_anne.

- What marital status you are: Avoid jane_single_female.

- What faith you are: Avoid muslim_jane.

- Where you live: Avoid jane_tampa_fl_usa.

- How old you are: Avoid fourteen_year_old_jane.

- What ethnicity you are: Avoid chinese_jane.

- Where you work: Avoid jane_xyz_corp.

- What you do: Avoid jane_the_model.

- How wealthy you are: Avoid loads_of_money_jane.

By following these guidelines you will end up with a Skype user name that is, shall we say, "robust" from a privacy point of view. But remember, it is for you to decide on the trade-off between what's useful to you and something that meets an arbitrarily high privacy standard.

To construct a password that is as "robust" as your Skype user name, follow these rules ...

- Don't use dictionary words (English or foreign): Avoid red_white_ and_blue. Use thgnbrtsbf.

- Don't use short passwords: Avoid red. Use long_passsswoooords.

- Don't use any personal information: Avoid AutoReg_123ABC. Use nothing_about_me_thgnbrtsbf.

- Do use mixed upper and lower case: Avoid all_lower_case. Use Mixed_LoweR_AnD_UPPer_CaSe_ThgnbrtsbF.

- Do use numbers: Avoid all_characters. Use Characters_AnD_ D1g1t5.

- Do use non-alphanumeric characters: Avoid LettersOnly. Use _Letter5_AnD_D1g1t5_.

- Don't use anything that can be easily guessed, even by people who know you well: Use passwords that are really, really hard to guess, but which you will be able to remember.

- Do change your password from time to time: How often you should change your password (daily, monthly, yearly, or whatever) is up to you and your level of paranoia! But passwords do grow old and less potent with age, especially if you end up using the password in many other applications besides Skype.

> ### Something Worth Knowing
>
> If you can't remember your password, don't despair! Provided that you gave a valid e-mail address when you created your Skype name, or added one to your online profile later, you can obtain a new password.
>
> From the Skype softphone sign in window, click on the link **Forgot your password?** or from the sign in page for the Skype website click on **Forgot your password?** Both actions will take you to the password reset web page on Skype's website. Follow the instructions, and Skype will reset your password and e-mail it to you.

Skype Privacy Settings

In the context of Skype, privacy involves controlling your visibility within the Skype online community and controlling the amount of personal information you share with others. What follows are the areas of Skype whose settings most directly impact your privacy when using Skype.

Automatic Change of Online Status

We covered the various settings for your online status in Chapter 10. However, there are a couple of additional settings in Skype that can affect your online status and are worth looking at.

You can have Skype automatically change your online status based on the length of time of inactivity of your keyboard and mouse on your PC. You can get to these settings by going to **Skype: Tools: Options ...** and in the window that appears, clicking on the category **General.**

By entering a number opposite **Show me as 'Away' when I am inactive for [fill in the number] minutes,** your online status will switch to Away after no keyboard or mouse activity on your PC for the amount of time you specify. Similarly, by entering a number opposite **Show me as 'Not Available' when I am inactive for [fill in the number] minutes,** your online status will switch to Not Available after the specific amount of time. Setting the number to zero for either of these two settings will leave your online status unaltered by any length of inactivity.

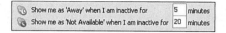

You can have Skype auto-matically change your online status based on the length of time of inactivity of your keyboard and mouse on your PC.

Auto-sign-in also affects your online status.

Auto-Start and Auto-Sign-In

At the Skype sign in screen (see the following figure), you can elect to have Skype start automatically when Windows starts and also sign in to your Skype account automatically. If other people have access to your PC and Windows user account, these auto-start and auto-sign-in features pose a risk.

At the Skype sign in window you can enable or disable auto-start and auto-sign-in for Skype.

You can get to the Skype sign in window by signing out (**Skype: File: Sign Out**). You can also change the auto-start Skype option from the Skype Options window (**Skype: Tools: Options ...** and click on the **Advanced** category, and then add or remove the check mark against **Start Skype when I start Windows**).

Block Unwanted Contacts

If you receive unwanted and persistent calls, chats, or authoriza-tion requests from a Skype user, you can block that user from ever again contacting you via Skype. To block a user, or indeed unblock a user previously blocked, open the Manage Blocked Users window (**Skype: Tools: Options ...** and in the window that appears, click on the **Privacy** category and then click on the link **Manage Blocked Contacts**), as shown in the next figure.

You can block users from ever being able to contact you via Skype.

To block a user, enter his or her Skype user name and click on the **Block User** button. To unblock a user, click on his or her Skype user name in your list of blocked users and click on the **Unblock User** but-ton.

You can also block a user who is currently in your contacts list without having to open the Manage Blocked Users window. Go to your contacts list, right-click on the user, and from the popup menu that appears, choose **Block This User.**

Your Online Profile

Chapters 10 and 11 walked you through the process of setting up your online profile. Those chapters show you how to change various entries in your profile. In this chapter we focus on what to put into, and espe-cially what to leave out of, your profile.

Settings for your profile are grouped into three categories:

♦ **Public details:** This category includes such things as your full name, gender, birth date, and contact details; and it is made available for all Skype users to see.

♦ **Contacts-only details:** Information in this category is only available to Skype users who are in your contacts list. It includes your picture, your local time, and how many contacts you have.

♦ **Private details:** You can add an e-mail address (in fact, up to three e-mail addresses) to your online profile. E-mail addresses are not visible to any Skype users. However, people who know one or more of your e-mail addresses can use them to search for you in the Skype online directory of users.

Caution!

Remember that Skype user names last forever. So once a Skype user name is created, it is never recycled and lingers on the Skype network forever. So if you switch user names, it is always a good idea to remove your personal details from your old public profile. It's not possible to actually delete your profile, but updating it with nonsense data will achieve the same effect.

Skype Video Options

If you have a webcam that works with Skype, there are a few privacy settings worth looking at, as shown in the following figure. You can open this window by going to **Skype: Tools: Options ...** and in the window that appears, clicking on the **Video** category.

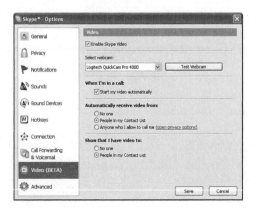

Settings for Skype video.

In terms of privacy settings, here's what you should look at:

◆ **Enable Skype video:** Remove the check mark against this item if you want to disable Skype video.

◆ **When I'm in a call start my video automatically:** If you answer an incoming call with this setting enabled, your camera will come to life, and whatever it sees will be broadcast to the caller. If you're in the habit of working at your computer in the buff, this might prove rather embarrassing! Remove the check mark against this item if you want to control your video manually.

◆ **Automatically receive video calls from:** The options are no one, people in my contacts list, and anyone.

◆ **Show that I have video to:** The two options here are no one, and people in my contacts list.

Skype Privacy Options

You can view and set Skype privacy options by going to **Skype: Tools: Options ...** and in the window that appears, clicking on the **Privacy** category, as shown in the following figure.

Skype privacy settings.

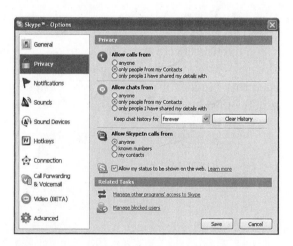

Most of the settings in the Skype privacy options window are self-explanatory. However, a few warrant additional explanation:

♦ **Keep chat history for:** You can choose to archive your chat session for a few different lengths of time, including forever, a set period of time (three months, one month, or two weeks), or not at all.

♦ **Allow my status to be shown on the web:** This feature allows you to control your visibility on the web through your Skype buttons (see Chapter 12), which are embedded in web pages or e-mails.

♦ **Manage other programs' access to Skype:** You can manage the access of add-ons to Skype by following this link. This issue is covered in detail in Chapter 16.

Low Privacy Profile

To configure Skype with fairly minimal privacy settings, use the settings in the following table.

Skype Settings for Low Privacy Profiles

Privacy Setting	Set To
Profile: Full name	Anything
Switch to: "Away"	5 minutes
Switch to: "Not Available"	20 minutes
Auto: Start Skype	Yes
Auto: Sign-in to Skype	Yes
Profile: Public details	Yes
Profile: Contacts-only details	Yes
Profile: Private details	Yes
Video: Enable	Yes
Video: Start automatically in call	Yes
Video: Receive calls from	Anyone
Video: Show you have video to	People in contact list
Privacy: Allow calls from	Anyone
Privacy: Allow chats from	Anyone

continued

Skype Settings for Low Privacy Profiles (continued)

Privacy Setting	Set To
Privacy: Keep chats	Forever
Privacy: Allow SkypeIn calls	Anyone
Privacy: Show online status	Yes

Medium Privacy Profile

To configure Skype with intermediate privacy settings, use the settings in the following table.

Skype Settings for Medium Privacy Profiles

Privacy Setting	Set To
Profile: Full name	Avoid names with gender, etc.
Switch to: "Away"	15 minutes
Switch to: "Not Available"	40 minutes
Auto: Start Skype	Yes
Auto: Sign in to Skype	No
Profile: Public details	Selective information only
Profile: Contacts-only details	Picture and local time
Profile: Private details	Yes
Video: Enable	Yes
Video: Start automatically in call	No
Video: Receive calls from	People in contact list
Video: Show you have video to	People in contact list
Privacy: Allow calls from	Contacts only
Privacy: Allow chats from	Contacts only
Privacy: Keep chats	1 month
Privacy: Allow SkypeIn calls	Contacts
Privacy: Show online status	Yes

High Privacy Profile

To configure Skype with restrictive privacy settings, use the settings in the following table.

Skype Settings for High Privacy Profiles

Privacy Setting	Set To
Profile: Full name	Name with little info in it
Switch to: "Away"	0 minutes
Switch to: "Not Available"	0 minutes
Auto: Start Skype	No
Auto: Sign in to Skype	No
Profile: Public details	Very selective
Profile: Contacts-only details	No (company logo acceptable)
Profile: Private details	No (company e-mail only)
Video: Enable	Disable
Video: Start automatically in call	N/A
Video: Receive calls from	N/A
Video: Show you have video to	N/A
Privacy: Allow calls from	People with shared details
Privacy: Allow chats from	People with shared details
Privacy: Keep chats	1 week
Privacy: Allow SkypeIn calls	Known numbers only
Privacy: Show online status	No

The Least You Need to Know

◆ Your privacy settings for Skype can influence your enjoyment of Skype, so it's worth getting them right.

◆ Choosing a sensible user name and robust password is the first step to obtaining your desired level of privacy when using Skype.

- Several Skype settings are important from a privacy point of view.

- By setting your privacy options according to one of the profiles provided, you can quickly and easily achieve your desired level of privacy.

Chapter 16

Managing Your Security

In This Chapter

- Avoiding viruses when transferring files
- Using Skype Add-on Security Management
- Following additional Skype security tips
- Adjusting your security settings to match for your security profile

The folks at Skype have been working hard to make their program secure. The Skype softphone comes with a lot of security features already built in. Even better, most of these features work behind the scenes! Even though a lot of the hard work for security has been done for you, there still remain some security risks that you, as the end user, must deal with. This chapter helps you manage this task.

Skype File Transfer and Viruses

As you know, you can use Skype to send files to other Skype users, just as they can send files to you. Unless you've been living in a cave for the past several years, you already know that the problem with receiving files from any source (including via e-mail and downloads from the Internet) is the possibility that such files might contain a *virus*.

def•i•ni•tion

A computer **virus** is a self-replicating and self-propagating program that can do great harm to your PC. Indeed, viruses are designed to remain undetected until disaster has struck and your PC has in some way been harmed: software applications may no longer work, and your PC may cease to function properly or may limp along with greatly reduced performance.

Even files sent by people you know pose a danger. This is because, unknown to them, their PC might have already been infected by a virus, which they can then unintentionally pass on to you—but only if you accept a file and then open it.

When anyone tries to send you a file, Skype pops up a window, as shown in the next figure, that asks for your permission to accept and start transferring the file. If you click on the **Cancel** button, the file will be refused. To accept the file, click on the **Save As ...** button, which lets you save the file to your hard disk.

You must give your permission before a file can be downloaded to your computer.

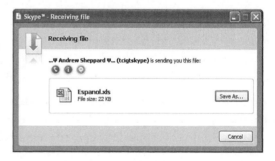

Something Worth Knowing

By default, Skype suggests that it save incoming files to this location: C:\Documents and Settings\[User name]\ My Documents\My Skype Received Files, where [User name] is the user name you use to sign on to Windows.

Regardless of who is sending you a file, you must first think long and hard before giving permission to receive it. Downloading files from people you know and trust poses only marginally less risk than from someone you don't know, because viruses do their most dastardly work when propagating in secret.

If you accept a file transfer from another Skype user, your first step once the transfer is complete—and certainly before opening and using the file—should be to scan it for possible infection by one or more viruses. Indeed, this is also true of any file you receive as an attachment to an e-mail or that you download from the Internet.

To do this you will need to use an antivirus software program. Most such programs scan your PC continuously, searching for infected files and fixing those it finds. The advantage of this type of antivirus program is that it automatically scans files as they are opened, so the act of opening a file also runs the antivirus program on that file. Alternatively, all antivirus programs allow you to scan a particular file in a specific location for viruses and, at the very least, it's this feature you should use to scan each and every file you receive from elsewhere.

The simple rule is that for any file you receive, via Skype or any other source, you must always first scan it for viruses before opening and using the file.

Skype Add-On Security Management

As you saw in Chapter 13, add-ons extend what Skype can do by establishing a two-way communication between another program and Skype. Clearly, were a malicious add-on to gain control of Skype, it could do some serious damage! For this reason, you must explicitly give permission for each and every add-on to use Skype.

When an add-on is first run and tries to establish communication with Skype, Skype will pop up an authorization window, as shown in the next figure, that gives you the following three choices:

- ◆ Allow this program to use Skype

- ◆ Allow this program to use Skype, but ask again in the future

- ◆ Do not allow this program to use Skype

Selecting the last of these choices will block the add-on from using Skype.

*You must explicitly give
your permission for each and
every add-on to use Skype.*

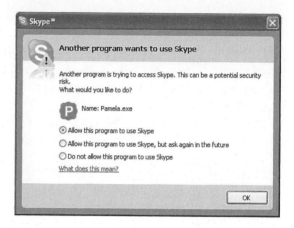

Should you wish to later change the access control for a particular add-on, you can do so by opening the Manage API Access Control window (**Skype: Tools: Options ...** and in the window that opens, click on the **Privacy** category, and then click on the link **Manage other programs' access to Skype**), as shown in the following figure. Simply select the add-on by clicking on its entry in the list of add-ons, and then click on the **Change** button. This pops open the Manage API Access Control window (see the previous figure) with the same three choices as before.

*Skype's access control window
for add-ons.*

Additional Skype Security Tips

There are a variety of additional actions you can take that further improve your security when using Skype.

Skype Updates

When a security vulnerability has been found for the Skype softphone, Skype is very quick to fix the program and issue an updated version of its softphone. By putting a check mark opposite **Check for updates automatically** in the Advanced category of the Skype Options window (**Skype: Tools: Options ...** and in the window that appears click on the **Advanced** category), you will be notified of both important new versions of Skype and any security updates. However, you will not be automatically notified of every minor revision to the Skype softphone, which often contain minor bug fixes and improvements.

In my experience, it always pays to be running the very latest version of the Skype softphone, so I suggest you check manually from time to time for new versions. You can do this by first checking what version you are presently running (**Skype: Help: About**), and then visiting the download page for the Skype softphone at www.skype.com/download/skype/windows. If there's a newer version, download and install it.

> **Caution!**
>
> Virus threats aren't exclusively targeted at Skype. Indeed, most are targeted at the Windows operating system and other Microsoft applications such as Internet Explorer. For that reason, you should first secure your PC by switching on automatic updates for Windows. To do this, go to **Start: Control Panel** and in the window that appears, double-click on **Automatic Updates.** This opens the **Automatic Updates** window; select **Automatic** and click on the **OK** button.

Authenticate Others

As you saw in Chapter 15, it's fairly easy to make an authorization request look genuine by using a known Skype user name with a "." (period) appended to it. I'm sure there are other ways in which people can try to impersonate others using Skype. Likewise, when you receive a phone call from someone who claims, say, to work at your company but whom you don't know, it makes sense to check their credentials in some way.

Authenticating the person at the other end of a Skype communication can be as simple as posing some simple questions for them to answer, or as concrete as carrying out a conference call with a third person who knows both the first and second parties very well.

Of course, the amount of effort you put into authenticating another person should be proportionate to how secure the communication needs to be. No one—but the totally paranoid—requires authentication of the pizza delivery guy!

Multiple Simultaneous Sign Ons

You can be signed on to the Skype softphone from several different computers simultaneously, which can be useful on occasion. For example, you may want incoming calls to ring simultaneously on both your home PC and your work PC; whichever picks up first, takes the call. This can be a faster and simpler alternative to call forwarding (see Chapter 7).

But it also means that someone could be signed into Skype somewhere under your user name at the same time you are, and you would have no idea! However, if you change your password (**Skype: File: My Skype Account: Change Password ...**), all other instances of your user name—wherever they might be—are automatically signed off within one hour or less (and immediately signed off upon trying to start a call, chat, or file transfer). Anyone wanting to use Skype under your name would have to sign in again, using your new password. By changing your password regularly, you can make sure that you, and only you, are signed on under your user name.

Secure Person-to-Person and Conference Calls

Skype-to-Skype calls, regardless of the number of Skype users involved, are secure because they're encrypted end-to-end. However, as soon as one party to a call is using a regular or mobile phone, the security drops dramatically because Skype's end-to-end encryption does not extend beyond its network. If you want to make secure calls, chats, or file transfers, make sure that everyone involved is using Skype.

Avoid Being a Supernode

In Chapter 1 you learned that some of the PCs that make up the Skype network are designated supernodes and take on extra functions. These extra functions consume both more of your computer's processing power and more network bandwidth. This can affect both the performance of your PC and the network to which it is connected. If either is a problem, you might want to try and avoid having your PC become a supernode.

Skype doesn't publicize how a PC running Skype becomes a supernode, but there's reliable evidence that disabling Skype's use of ports 80 and 443 (**Skype: Tools: Options ...** and in the window that appears, click on the category **Connection** and make sure there's no check mark opposite **Use port 80 and 443 as alternatives for incoming connections**) stops your PC from becoming a supernode.

Phishing and Other Scams

Phishing is the act of tricking someone into giving away valuable information. You might receive a call, an e-mail, or a chat that asks you for information. Often such requests are worded carefully to lull you into lowering your guard. The really sophisticated ones direct you to genuine-looking, but fake, websites to try to persuade you to enter your personal details and information.

Skype will never ask you to disclose your password or other important information (such as credit card details) by means of a call, chat, or e-mail. If you ever receive such a request, don't trust it.

If you need to update or change information having to do with your Skype account, always do so through the official website at www.skype.com and no other.

Low Security Settings

To configure Skype with fairly minimal security settings, use the settings in the following table.

Some of the actions indicated in the following table, and the tables after that follow, involve adjusting settings on the Skype softphone; others are security precautions or actions you must take on your own. The latter are denoted with an asterisk (*).

Skype Settings for Low Security Profiles

Security Setting	Set To
Virus scan incoming files*	Always
Allow add-ons to use Skype	Yes
Automatically check for updates	No
Manually check for updates*	No
Windows automatic updates*	On
Authenticate others*	No
Change password	Occasionally
Secure calls (everyone using Skype)	No
Avoid becoming a supernode	No
Phishing and other scams*	Vigilance

Medium Security Settings

To configure Skype with intermediate security settings, use the settings in the following table.

Skype Settings for Medium Security Profiles

Security Setting	Set To
Virus scan incoming files*	Always
Allow add-ons to use Skype	Yes
Automatically check for updates	Yes
Manually check for updates*	Weekly
Windows automatic updates*	On
Authenticate others*	Only as needed
Change password	Monthly
Secure calls (everyone using Skype)	No
Avoid becoming a supernode	No
Phishing and other scams*	Vigilance

High Security Settings

To configure Skype with restrictive security settings, use the settings in the following table.

Skype Settings for High Security Profiles

Security Setting	Set To
Virus scan incoming files*	Always
Allow add-ons to use Skype	Yes, but always ask
Automatically check for updates	Yes
Manually check for updates*	Daily
Windows automatic updates*	On
Authenticate others*	Frequently
Change password	Weekly
Secure calls (everyone using Skype)	Yes
Avoid becoming a supernode	Yes
Phishing and other scams*	Extreme vigilance

The Least You Need to Know

 ◆ Skype has taken a lot of measures to ensure the security of its users.

 ◆ You should *always* scan all incoming file transfers for viruses.

 ◆ Not all the threats that affect Skype security come from Skype itself. You must also secure your PC.

 ◆ By setting your security options according to one of the profiles provided, you can quickly and easily achieve your desired level of security.

Part 5

Troubleshoot Skype

Skype boasts that it "just works." And indeed, for most Skype users most of the time, it hums along with nary a glitch. However, for those (hopefully rare) occasions on which Skype acts up, this section should prove invaluable. In these chapters I offer up simple, methodical procedures for finding and fixing most problems. Moreover, for those readers who lament the loss of 911, 411, and regular fax that comes from switching fully to Skype, I suggest tips and workarounds so that you can reclaim this lost functionality.

Chapter 17

What to Do When Things Go Wrong

In This Chapter

- ◆ Troubleshooting

- ◆ Consulting Skype user guides

- ◆ Using the Skype troubleshooter and knowledgebase

- ◆ Appealing to Skype user forums

- ◆ Sending support requests

Most users of the Skype softphone never encounter a problem, and I hope you remain among their number and never have to use the advice in this or the next two chapters. But, if you do have problems, these three chapters on troubleshooting Skype should help.

Most Skype problems generally fall into two different areas: networking problems and audio and video problems. If you are going to have a problem, odds are that it will be in one of these two problem areas. Chapter 18 covers audio and video glitches, and Chapter 19 tackles problems associated with networking.

This chapter gives general troubleshooting strategies and advice, and together with Chapters 18 and 19, should help with the vast majority of problems you are likely to encounter when using Skype. Good luck!

Something to Try

A methodical approach to fixing problems with Skype takes time. Meanwhile, if you desperately need to use Skype, consider using another computer until your problem is fixed. That way, you'll reduce the pressure to fix Skype on your primary PC.

Troubleshooting Strategies

If I could put a bold slogan across this troubleshooting section of the book, it would read: *Don't Panic!* It's good advice. Launching yourself frantically at the problem can often do more harm than good. So, stay calm, have a cup of tea, and be methodical about fixing the problem.

General Advice on Troubleshooting

Think of a problem with Skype as a wild beast. It helps to drive the thing into a corner before you kill it!

The way to drive a problem into a corner, so you know exactly what the problem is and where it is, and so have a better chance of fixing it, is to be systematic and methodical. In a notebook, keep track of the things you have tried and their results. Without systematic note taking, you face the risk of going in circles and repeating tests to which you already knew the answer.

Updates

If you're having problems with Skype, the very first thing you should do is make sure you are running the latest version. Open Skype and go to **Skype: Help: About** to see what version you are currently running on your PC. Next go to www.skype.com/download/skype/windows to see what the most up-to-date version of Skype is available for download. If there's a newer version available, download it and install.

Also look at the change log for the latest version of Skype at www. skype.com/download/skype/windows/changelog.html to see whether

the problem you are having has
already been fixed in the latest
version of the Skype softphone.
Software problems are called *bugs*,
and the change log lists what bugs
have been fixed from previous versions.

If downloading the latest version of
Skype and installing it fixes your
problem, great, you can stop right
here! If not, read on.

def•i•ni•tion

Problems with software programs are known as **bugs**
because in the first computers,
which used old vacuum tubes,
insects sometimes caused problems. Nowadays, **bug** is a
catch-all term for anything that
goes wrong with a program.

Skype User Guides

If you're having problems with Skype, particularly if you're new to
the program, you should first check to see whether you are using it
properly. Take a few minutes to review the basics of how to use Skype
by visiting the screenshots user guide at www.skype.com/download/
screenshots.html, which is shown in the following figure.

*Online screenshots and
instructions for use with the
Skype softphone.*

For more detailed instruction on how to use specific features of Skype,
consult the online user guides at www.skype.com/help/guides.

The Skype Troubleshooter

The Skype online troubleshooter is a handy tool that helps users solve basic problems, as shown in the next figure. By answering a series of simple questions, the troubleshooter directs you to the most appropriate solution to your problem. The Skype troubleshooter can be found at support.skype.com/?_a=troubleshooter.

Skype's online troubleshooter helps you fix problems step by step.

Because of its rigid question-and-answer format, the Skype troubleshooter can't solve complex problems for you. For those, you'll need to turn to more advanced tools, such as Skype knowledgebase or online forums.

The Skype Knowledgebase

Skype's online *knowledgebase* is a compilation of answers generated in response to user support requests and questions. It not only contains information on a great number of topics of general interest to Skype users, it also contains advice on how to fix problems with Skype. One tremendous advantage of the knowledgebase over what has been described so far in this chapter is that it's searchable. The Skype online knowledgebase can be found at support.skype.com/?_a=knowledgebase.

For example, suppose you are sending a file. It's transferring unusually slowly, and you see the message "Your transfer is being relayed" in the file transfer window. If you don't already know what "relayed transfer" is, enter it as a search in Skype's knowledgebase, and you will get the answer (in fact, several answers in varying degrees of technical detail).

def•i•ni•tion

A **knowledgebase** is a database of information about a particular topic, structured in such a way that it's searchable.

The point is that many of the non-basic features of Skype are not documented anywhere else other than the knowledgebase.

Search Skype's online knowledgebase for problem fixes and other information.

The Skype User Forums

Many people use Skype's online forums to exchange information. You shouldn't hesitate to consult the forums, as many of the people who participate in them are very knowledgeable about Skype. There are other places on the Internet to seek advice about Skype (see Appendix B), but Skype's user forums are the most popular. Skype's user forums can be found at forum.skype.com.

Messages posted to the forums are grouped into different categories, as shown in the following figure.

Skype's online user forums contain a wealth of information.

Anyone can browse and search the forums, but you can only post messages and questions if you first register. Registration is free and requires only an e-mail address in addition to choosing a user name and password.

Search the Skype Forums for Answers

An indispensable feature of the Skype forums is that they can be searched, as shown in the following figure. Using a combination of search terms you can narrow the hundreds of thousands of posted messages down to a manageable handful relevant to the problem you're trying to fix.

Search Skype's online user forums to find answers to specific problems.

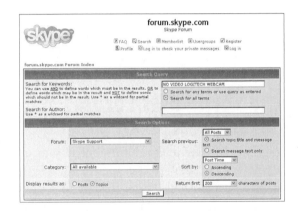

Post Questions to the Skype Forums

If a search of the Skype forums doesn't turn up an answer, you can always post a question to the forums regarding your problem. You are almost certain to get an intelligent answer within 24 hours, usually much faster.

Support Requests

If none of the other tools recommended in this chapter has helped you solve your problem, you can submit a support request. Visit www.skype.com, and from the menu at the top of the home page, click on **Help.** This will take you to the Getting Help for Skype web page, and by clicking on the **Support Requests** link you will be presented with a support request form, as shown in the next figure.

Submit a Support Request

Before you submit a request please be sure you have read through our Knowledgebase and User Guides. If you can't seem to find a solution to your problems with Skype in our knowledgebase or our our troubleshooter, you can fill out the fields below with as detailed information as you can and send it to us. * Indicates a required field.

Please only submit your support request in **English, French, German** or **Spanish**

Name*

Email*

Subject*

Skype Name*

Skype Version

Message*

Department
Billing

Attach a file

Browse...

Submit the support request

Submit a support request directly to Skype.

If you're lucky, you might get a reply from a member of Skype's technical staff. However, don't bank on it. I have never received a reply to any support request I have submitted, though I do know a handful of lucky people who have. This situation might improve if, and when, Skype puts more resources into support. In the meantime, let us hope you are one of the lucky ones who gets a reply!

Something Worth Knowing

If you are having problems specifically with a SkypeOut call, it might be Skype, or it might be a problem where Skype joins the public telephone network. Consequently, Skype has a reporting service just for this type of problem. Send an e-mail to pstn-feedback@skype.net and include in your message a description of the problem, the number (including country code) you were trying to call, your location, the time of day and time zone, and your Skype name.

The Least You Need to Know

- ◆ To troubleshoot Skype effectively, you must approach the problem calmly and systematically.

- ◆ Make sure you have the most up-to-date version of the Skype softphone.

- ◆ You should first consult Skype's user guides to make sure you are using the program correctly.

- ◆ You can use Skype's troubleshooter to find answers to basic problems.

- ◆ You can search Skype's knowledgebase and online user forums to find solutions to more complex problems.

18

Troubleshooting Skype Audio and Video

In This Chapter

♦ Reviewing your Skype audio and video settings

♦ Testing Skype audio

♦ Testing Skype video

♦ Decoding Skype audio and video error messages

Skype-to-Skype calls typically have very good audio voice quality—better than the quality for most regular and mobile phones, in fact. SkypeOut calls have lower voice quality, but are still quite good. Video quality is likewise very good. However, Skype audio and video can go awry, and it's the purpose of this chapter to help you fix them when they do.

Skype Audio and Video Options

If you're having problems with either Skype audio or video, you should first check your option settings to see whether anything has changed or whether the settings you still have are appropriate. For example, if you've installed new hardware, you might need to update your Sound Devices settings. Option settings for audio and video are covered in Chapters 2, 4, and 10.

Test Skype Audio

If the problem persists after checking your settings, it's time to start isolating the problem. You must determine whether your audio problem is caused by Skype or by a hardware device or another piece of software on which Skype depends.

Test Your PC Audio Independently of Skype

By first testing your PC audio system independently of Skype, you can decide whether the problem is with Skype or your PC.

To test your PC's audio system independently of Skype, try the following tests:

- **To test your computer's audio-out component:** Open the Windows Sound Recorder application (**Start: All Programs: Accessories: Entertainment: Sound Recorder**), which is shown in the following figure. With Sound Recorder open, select **File: Open** and then navigate to **C:\ Windows\Media.** Open any .wav file (that is, any file with a .wav extension) you find there and click the **Play** button (the one with a single black right-pointing arrow on it). If you hear a sound, the audio-out component of your PC sound system is working.

> **Something Worth Knowing**
>
> Skype provides a user guide specifically for Windows XP audio setup. It can be found at www.skype.com/help/guides/soundsetup_xp.html.

Note that you can check which audio devices are associated with sound recording and sound playback by going to **Edit: Audio Properties** in Sound Recorder.

- **To test your computer's audio-in component:** With Sound Recorder open, click on the **Record** button (the one with a red dot on it) and speak into your audio-in device for a few seconds.

Click the **Stop** button (the one with a black rectangle on it) and then click the **Play** button. If you hear your voice played back, the audio-in component of your PC sound system is working.

The Windows Sound Recorder application.

An alternative method for checking that your PC sound system is working is to run the Windows Sound Hardware Test Wizard. Go to **Start: Control Panel** and in the window that appears, double-click on **Sounds and Audio Devices;** this opens the Sounds and Audio Devices Properties window. Next navigate to the **Voice** tab and click on the **Test Hardware ...** button, which opens the Sound Hardware Test Wizard, shown in the next figure. Follow the instructions of the wizard to test your PC sound hardware.

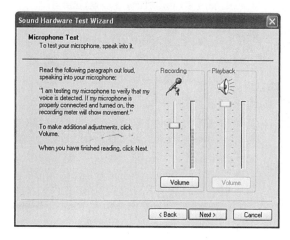

The Windows Sound Hardware Test Wizard.

If your PC audio system is working independently of Skype, chances are the problem is with Skype and not with your PC audio system.

> **Caution!**
>
> When Skype is first installed it uses the Windows pre-defined audio input and output devices for voice, called **Windows default device** in Skype's sound option settings. If you subsequently add new sound hardware to your PC, you will quite likely find that the default sound device for Windows, and therefore Skype, has changed. To avoid erroneous audio/ sound device selection issues in Skype, it's always best to explicitly specify the audio-in and audio-out devices for Skype in the sound devices options window (**Skype: Tools: Options ...** and in the window that opens, click on the category **Sound Devices**).

Test Your Audio Using Skype

Skype provides the user name echo123 for test call purposes. By default, this user appears as "Skype Test Call" in your contacts list.

echo123 is simply a Skype user name set up by Skype and linked to an automatic sound recording and playback system. Enter echo123 in the Skype address bar and start a call. A voice on the other end will guide you through the process of recording your own voice and then playing it back to you. If you can hear the automated announcement and your own voice when it is played back to you, you know that your audio-in and audio-out devices are working properly.

Something Worth Knowing
echo123 is an English-speaking service. The user name echo-chinese is a similar service for Chinese (dialect Mandarin, Taiwan) speakers; soundtestjap-anese is for Japanese speakers; and testyuyin is another voice test service for Chinese (dialect Mandarin, mainland China).

As a further test, you can have Skype call you! Open a chat session with echo123 and send the text message "callme" (without the quotes), and Skype's echo test service will call you back for a voice test.

Test Skype Video

The test method for video follows the same pattern as the tests for audio. The purpose of the video tests is first to isolate the problem, and then fix it. Try the following tests to determine whether you have a problem with Skype, or with your video webcam.

Test Your Video Independently of Skype

Most webcams come with software that makes the webcam work with your PC. Given the sheer variety of webcams available, I can't cover every make and model, but I can give you some pointers on the general approach using a specific example.

I use a Logitech QuickCam Pro 4000, and here's how to test it independently of Skype. Open the application for your webcam. In my case it's the Logitech QuickCam application (**Start: All Programs: Logitech: Logitech QuickCam**), which is shown in the following figure.

The Logitech QuickCam application.

If you see a picture from your webcam in the application that came with the webcam, you know it works. It's time to test your webcam with Skype.

Test Your Video Using Skype

Testing your webcam video using Skype was covered in Chapter 4, but I repeat the test here for your convenience.

To test whether your webcam is working with Skype, go to **Skype: Tools: Options ...** and in the window that appears, click on the **Video** category. This displays the video option settings for Skype. From the **Select webcam** pull-down menu, choose the webcam installed on your PC that you want to use with Skype, and then click on the **Test Webcam** button. This should open the Webcam Test window, as shown in the following figure.

Skype webcam test.

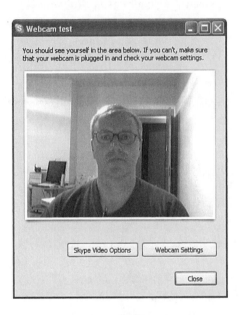

If you see a picture from your webcam in the Webcam Test window, you know it works with Skype.

If you are still having problems with video calls, you'll need to research the problem using the tools and techniques described in Chapter 17.

Skype Audio and Video Error Codes

Skype often notifies users of a problem by means of error codes and messages, as shown in the following figure.

Here is a list of the most common errors codes associated with audio problems for Skype:

♦ **Problem with playback device (#6101):** This error code means that Skype is unable to detect, or use, your audio-out device (speakers, headset, or handset). Try manually adjusting your audio-device settings in Skype (see Chapters 2 and 10). If that doesn't solve the problem, try downloading and installing the latest software drivers for your sound device. You can normally download drivers from the website of the device manufacturer. Finally, check to see whether there's a conflict between Skype and another application trying to use the same sound device. You can do this by closing all other applications except Skype and making a test call.

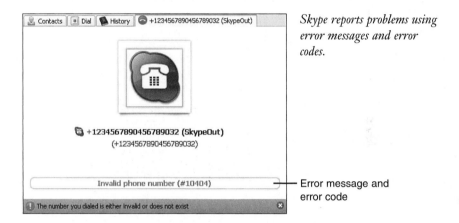

Skype reports problems using error messages and error codes.

Error message and error code

♦ **Problem with recording device (#6102):** This error code means that Skype is unable to detect, or use, your audio-in device (microphone, headset, or handset). Just as for error code 6101, the first two things you should try are: 1) manually adjusting your audio-in device settings in Skype; and 2) downloading and installing the latest software drivers for your sound device. Next, see whether there is a conflict between Skype and another device using your audio-in device by closing all applications other than Skype.

At the present, there are no error codes specific to Skype video. However, this might very well change in the future, so if you're having problems with video, search the Skype online knowledgebase for error codes specific to video.

The Least You Need to Know

♦ Audio is one of the most common problem areas for Skype users.

♦ If your audio or video stops working after having worked fine, you should first check your Skype option settings to make sure nothing has changed.

◆ Check your Skype option settings after you install new audio or video hardware on your PC, to make sure they are correct in Skype.

◆ By testing the audio and video systems of your PC independently of Skype, you can determine whether your problem is with Skype or with something else.

◆ You can use Skype error messages to help isolate and fix a problem.

Chapter **19**

Troubleshooting Skype Networking

In This Chapter

◆ Testing your network connection

◆ Decoding Skype networking error messages

◆ Making Skype work with firewalls

◆ Avoiding relayed file transfers

In many ways Skype's reputation for "Internet telephony that simply works" rests on the fact that the Skype softphone is network savvy. For most users, Skype self-configures its network settings. However, networking problems do occur, and this chapter is here to help you to troubleshoot them.

Test Your Network Connection

Problems with Skype networking, and especially with jittery or stuttering sound, might be due to network congestion or some other problem with your Internet connection that is unrelated to Skype. To test your network bandwidth and latency independently of Skype, you can use the same tests as those described in Chapter 2. If your network connection meets the bandwidth and latency criteria set forth in Chapter 2, it's probably okay, and you need to look elsewhere to identify and solve your problem.

Skype Networking Options

If you're having a problem with Skype's network connection, you should first review your network settings for Skype, as shown in the following figure. To open the Skype Options window for network connections, go to **Skype: Tools: Options ...** and in the window that appears, click on the **Connection** category.

Skype's network connection option settings.

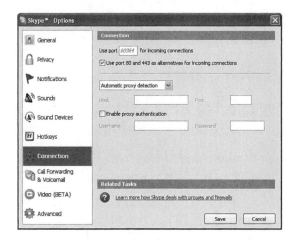

There are four parts to Skype's network settings, all of which you can adjust when trying to fix a problem (though some you should only experiment with under the guidance of a networking expert):

♦ **Use port XXXXX for incoming connections:** XXXXX is typically a 4- to 5-digit number, between 1024 and 65535, and is normally automatically assigned by Skype. It can be overridden, though this should only be done by someone, such as a network administrator, who thoroughly understands networking.

♦ **Use port 80 and 443 as alternatives for incoming connections:** By putting a check mark opposite this option, you should improve Skype's network connection, as you give Skype more flexibility. However, evidence suggests that doing so will increase the likelihood that your PC will become a supernode (for more on supernodes, see Chapters 1 and 16).

◆ **Automatic proxy detection:** This is the default setting, though the pull-down list has two other options: HTTPS and SOCKS5. Unless you already know what these additional options are, don't touch them unless under the guidance of a networking expert.

◆ **Enable proxy authentication:** If you connect to the Internet through a *proxy*, you may have been assigned a user name and password. Check the box and enter them here.

> **def•i•ni•tion**
>
> A **proxy** is a computer that is placed between your PC and the Internet, and whose purpose is to speed up, make more secure, and otherwise improve your interaction with the Internet.

Skype Networking Error Codes

As noted in Chapter 18, Skype often notifies users of a problem by means of error codes and messages. Here is a list of the most common errors associated with networking problems for Skype:

◆ **Error message "Could not connect to Skype proxy":** You get this error message when your SkypeOut call is unable to be completed because one of Skype's network gateways has reached its capacity. Try again and your call should be routed through a gateway that hasn't reached its call capacity.

◆ **Error codes "1101/1102/1103":** These error codes indicate that your Skype softphone was unable to connect to the Skype network. Test your Internet connection to make sure it's still working, and check that your network settings (and firewall settings, see the next section) haven't changed.

◆ **Error codes "6503/6504/10500/10503":** These error codes are typically associated with the "Service Unavailable" error message, which usually means that the Skype network is temporarily overloaded. Wait 15 to 30 minutes and try again.

◆ **Error code "10408":** This is the error code for "Unable to Connect—Call Timed Out" and occurs when a call is unable to connect with the other party within a reasonable amount of time. Wait 15 to 30 minutes and try calling again.

Making Skype Work with Firewalls

Most PCs connect to the Internet through a *firewall* for security purposes. Firewalls can take the form of software or hardware. A software firewall normally runs on your PC as just another application, whereas hardware firewalls are dedicated computers or are built into hardware that sits on your network. Both software and hardware firewalls can cause problems for Skype.

def•i•ni•tion

A **firewall** is a gatekeeper for your Internet connection. It blocks malicious network traffic from reaching your computer, thereby providing a more secure Internet experience.

Skype has a special page on its website to show users how to configure Skype to work around problems with firewalls. It can be found at www.skype.com/help/guides/firewall.html.

Something to Try

Most modern network hardware, such as broadband modems and routers, have on-board software (that sometimes includes a firewall) that you can update. This is normally referred to as firmware. It's always worth updating the firmware for your network hardware to the latest version, as that can sometimes cure a number of problems and glitches that impact Skype. Visit the website for your network hardware manufacturer and see whether there's a firmware update you can download and install on your hardware.

Relayed File Transfer

Skype enters relayed file transfer mode when the softphone is unable to make a direct point-to-point connection between your PC and the recipient's PC. Relayed file transfers are substantially slower than direct file transfers, because data is relayed via other nodes in the Skype network and, as a consequence, Skype throttles back the data-transfer speed to a maximum of only 1 kilobyte per second. To put this into perspective, a 1 megabyte file will take a minimum of 1,000 seconds (approximately 17 minutes) to send via relayed transfer, whereas a direct point-to-point transfer will probably take less than a minute! Clearly, relayed transfer is to be avoided, if at all possible.

You are notified of a relayed transfer by a message to that effect displayed in the lower-left corner of the file-transfer window, as shown in the following figure.

You are notified by Skype when a file transfer is relayed.

Relayed file transfer

Relayed file transfer is a function of both your network connection and that of the file recipient. To better improve your chances of avoiding relayed file transfer, both you and the file recipient should try the following:

- ◆ Try an alternative machine on a different Internet connection. Granted, this isn't an option for many people, but for those who do have this option, it is something quick and simple to try.

- ◆ Kill all chat sessions. There's some evidence that having open chat sessions will sometimes put file transfer into relay mode.

- ◆ Try changing your network router and firewall settings. Specifically, open up more network ports to Skype and make your Internet connection settings more open and generous. Specifically, switch off Network Address Translation (NAT), and open all ports above 1024 to unrestricted incoming and outgoing UDP and TCP data packets. If that sounds like gobbledygook to you, you'll need the help of someone knowledgeable about networking.

Another reason to try and improve relayed file transfer speed is that voice calls can also be relayed, and the more relay points between you and a call recipient the more likely that call quality will suffer as a result.

The Least You Need to Know

◆ Skype is, by and large, self configuring from a networking point of view.

◆ If you're experiencing network problems with Skype, first check your Internet connection independently of Skype.

◆ You can use error messages and error codes from Skype to identify and fix a problem.

◆ Problems with firewalls and network routers can force Skype into "relayed" mode, which degrades file transfer and voice calls.

Chapter 20

Workarounds for 911, 411, and Fax

In This Chapter

♦ Establishing workarounds for 911 emergency services

♦ Establishing workarounds for 411 directory services

♦ Establishing workarounds for fax

Skype doesn't provide access to 911 emergency services or 411 directory assistance services. Nor does it support regular fax. Period.

Nevertheless, there are some quite good workarounds for these services that enable you to reclaim most of this lost functionality. These workarounds aren't perfect, and you must use them solely at your own risk, but they are nevertheless workable alternatives for most Skype users.

Workarounds for 911 Emergency Services

If you dial 911 in Skype, you will not be connected to any emergency services. Skype itself is quite explicit on this issue, and therefore markets its services as an "enhancement" to regular telephone services.

What follows are some workarounds on how to create a "911-like" service. Perhaps it is worth repeating that the workarounds are only "911-like" and are in no way a direct substitute for real 911 services.

Caution!

Your home alarm system is probably remotely monitored through your telephone line. So if you are going to keep at least one regular phone line, make sure it's the one used by your alarm system. If you decide to use Skype exclusively, you will have to give up remote alarm system monitoring. (This isn't strictly true, as you can modify your home phone wiring to work with your alarm system using Skype, but that requires a high degree of technical and do-it-yourself skills, and so isn't discussed in this book.)

You May Have 911 Service by Default

Most people get their broadband Internet service through one of two means: cable TV or telephone. Cable Internet uses the same physical cable through which you receive your TV channels, so it is independent of your telephone service. You can have cable Internet access without having a telephone, and vice versa.

Telephone can also provide broadband Internet access, usually at an additional cost to your regular telephone services. The most common technology for providing Internet access through your telephone line is a digital subscriber line (DSL). There are many flavors of DSL, most of which are bundled with and ride atop a regular telephone service. If you obtain your Internet connection through a flavor of DSL that rides atop a regular telephone service, you already have real 911 services as part of your regular telephone service. Thus, if you have ADSL or another flavor of DSL that piggybacks on top of your regular telephone service, you will have 911 emergency services by default.

Getting your broadband Internet connection through DSL with telephone services does not stop you from switching to Skype in large part to save money. Your local telephone company might not promote lower-cost service packages; indeed, it may not mention them at all! However, most telephone companies do offer "basic" or "emergency-only" service at much lower cost than their more visible offerings. In the case of "emergency-only" service, the cost may be as low as $10 to $20 per month. Depending on your local telephone company, this may be combined with DSL to reduce your regular telephone costs to the bare minimum, leaving Skype as your principal low-cost telephone provider, but retaining normal access to 911 emergency services.

Set Up a "911-Like" Service

If you obtain your Internet connection through some mechanism other than ADSL (cable, say) you can switch to Skype as your sole telephone service provider.

As you learned in Chapter 5, SkypeOut is a paid service that enables you to call regular telephones. Many local police departments and fire services have telephone numbers separate from the 911 service. Moreover, these numbers are often (although not always, so please check) staffed 24-hours a day. In an emergency, you can call your local police or fire service using SkypeOut, and your request for help will surely be heeded.

As a matter of convenience, you should add your local police station to your contacts list in Skype and assign it a speed-dial number that you can easily remember. In addition, you may want to add a fast-dial shortcut to your desktop (see Chapter 22) for quickly dialing your local police department in an emergency—when seconds might count! Once setup, don't forget to test both your speed-dial and fast-dial shortcuts.

Use a Mobile Phone as a Backup

All U.S. mobile phones support 911 services. In addition, the Federal Communications Commission (FCC) mandates that all mobile phones, regardless of the status of service (or even the lack of service), must be able to dial 911. Whether you have a telephone number or not, or whether you have signed up for service or not, a mobile phone can connect to 911. So, if your home is in an area with good mobile phone reception, it's always worth keeping your mobile phone charged and ready should you need it in an emergency. Indeed, you can even use an old mobile phone that is no longer supported by a paid calling plan from a mobile phone company.

> ### Something Worth Knowing
>
> One clear advantage of regular 911 emergency services is that when you place an emergency call from a regular phone, the dispatcher knows where you are located. This is not the case when you place any sort of emergency call using SkypeOut, as the nature of Skype's technology means that your location is unknown. However, if you use a mobile phone to place a call to 911 emergency service, in many instances the dispatcher will know where you are (at least approximately), as many mobile phone networks have built-in location detection. It's no guarantee, but it's something worth knowing!

Workarounds for 411 Directory Services

In the United States, you can access directory assistance by dialing 411. This is usually a paid service, and $1.25 per directory assistance call is not untypical. Skype does not support 411 directory assistance services. Dial 411 and you'll get nothing but an error message. That's the good news, as you'll never have to pay for 411 calls again!

A free alternative is to use one of the many online 411 services that are now available. A search on Google with the keywords "411 directory assistance" will spit out a long list of such directory assistance websites. But perhaps the first place to go is the list of online directory assistance services maintained by the Telecommunications Research & Action Center (TRAC), a nonprofit watchdog that promotes the interests of residential telecommunications consumers. Given the charter under which TRAC operates, the services listed there are among the less annoying websites for getting directory assistance. The website for TRAC is www.trac.org; navigating to the directory assistance page is easy from there.

Workarounds for Fax

Fax simply doesn't work over a Skype connection, regardless of whether you use a software or a hardware fax device.

The reason why fax doesn't work with Skype is too complex to go into here, so just take my word for it: at the present time, Skype doesn't work with regular fax machines.

Fortunately, you have a lot of choices when it comes to workarounds for Skype's lack of support for fax.

Keep Your Fax Machine

Switching to Skype is not an all-or-nothing proposition. You can choose to mix and match Skype services with other Internet telephone offerings or even with your *POTS*. If you retain one or more regular telephone lines, you can connect your fax to one of them to sidestep the problem altogether.

def•i•ni•tion

Plain old telephone system, or POTS, is a term referring to the services provided by your regular telephone system.

Use Skype File Transfer Instead

If the document is in electronic form, instead of printing it and faxing it, consider sending it via Skype's file-transfer function. Of course, this will work only if your recipient is also running Skype. A plus in favor of this workaround is that it typically costs nothing to send files via Skype, whereas regular fax costs money.

Online Fax Services

Enter the search term "fax service" in an Internet search engine and you'll be presented with a plethora of different types of fax services that don't require you to have a regular fax machine to send or receive faxes. Some of these services are free, but most are available at low cost. It is a competitive market segment; which is, of course, to your benefit.

Consider Alternative VoIP Providers

Some VoIP providers do provide fax-compatible services, though usually at an additional cost. You might want to consider one of these providers instead of, or as a complement to, Skype.

The Least You Need to Know

- Skype does not support 911, 411, or regular fax machines.

- There are a number of workarounds, which you must use at your own risk, that reclaim much of this lost functionality.

- Keeping at least one regular phone line means you will have access to genuine 911 calling.

- Mobile phones with, or without, an active subscriber plan work for placing 911 calls.

- 411 directory assistance services are easily obtained through free alternatives.

- Although your regular fax machine will not work with Skype, there are alternatives, many of which are free.

Part 6

Accessibility and Usability

Accessibility and usability are the concerns of everyone—not just those with disabilities. Making Skype easier to use makes it a more productive tool while also making it that much more enjoyable. Use this section to simplify and accelerate how you use Skype.

Chapter 21

Making Skype More Accessible

In This Chapter

- ◆ Making Windows more accessible
- ◆ Increasing Skype's accessibility
- ◆ Using your voice to operate Skype

Usability is the simplicity and efficiency with which something can be used, whereas accessibility is making something accessible to those with some kind of visual, hearing, or dexterity impediment. All users of Skype should be concerned with both usability and accessibility, because they affect our productivity as well as our pleasure.

Even for those of us with no physical impediments, accessibility features such as operating Skype using only our voice, are clearly of great benefit—anyone who has a PC in the kitchen can attest to that! Moreover, nobody is forever young, so all of us—one day—might find our motor skills, or eyesight, or hearing, or all three, are not quite as acute as they once were. That is, everyone has a vested interest in *assistive technologies*.

def•i•ni•tion

Assistive technologies
describes hardware and software that enable people with impaired abilities to make easier and better use of something. For Skype, this mostly comes down to making its graphical user interface more usable and accessible.

Unfortunately, even though Skype has made great strides in the areas of usability and accessibility, there are still a few gotchas. No doubt Skype will continue to make improvements, but in the meantime there are quite a few things you can do to improve both the usability and accessibility of the Skype softphone. This chapter and the next show you how.

Make Windows More Accessible and Usable

By making Windows itself more usable and accessible you will, in turn, make Skype (and the other applications you use) simpler and easier to use. What's even better is that these extra features for Windows are built in—all you have to do is turn them on.

Something Worth Knowing

To learn more about how you can configure Windows to be more accessible, visit the Microsoft Accessibility website at www.microsoft.com/enable.

Windows comes with a number of accessibility features and utilities built-in, including the following:

◆ **Accessibility Wizard** (Start: All Programs: Accessories: Accessibility: Accessibility Wizard): This wizard will take you step by step through configuring Windows on your PC so that it is more accessible to those with special needs.

◆ **Magnifier** (Start: All Programs: Accessories: Accessibility: Magnifier): This is a utility that makes things on your computer screen more readable by magnifying a selected portion of the screen.

◆ **Narrator** (Start: All Programs: Accessories: Accessibility: Narrator): This text-to-speech utility will read out loud what appears on your screen. It can also speak what you type on the keyboard.

◆ **On-Screen Keyboard** (Start: All Programs: Accessories: Accessibility: On-Screen Keyboard): This is a virtual keyboard for your computer screen that enables you to emulate keyboard input using only your mouse.

◆ **Utility Manager** (Start: All Programs: Accessories: Accessibility: Utility Manager): This feature is used to manage the accessibility utilities. Note that you must have administration privileges on your PC to make full use of the Utility Manager.

From the point of view of Skype, the Accessibility Wizard and Narrator are the two most useful accessibility utilities from the previous list. For example, if Narrator is on, when you select a contact in your list, Narrator will read the name of the contact out loud.

In terms of both usability and accessibility for Windows in general, and Skype in particular, you might also want to revisit Chapter 10, which shows you how you can change the look and feel of the Windows and Skype user interfaces to your advantage. Likewise, the next chapter, Chapter 22, shows you how to drive the Skype user interface using only your keyboard, which can often be far faster than using the mouse; and, in the case of the visually impaired, it's a practical alternative to using the mouse.

Speech Recognition Programs and Other Accessibility Tools

Skype is such a new company that there hasn't been enough time for a third-party market to form for add-ons that improve its accessibility. However, there is at least one accessibility program for Windows that has been customized to work with Skype.

JAWS (short for Job Access With Speech; www.freedomscientific.com) is an application that reads information that appears on your screen using a synthesized voice (similar to Microsoft's Narrator utility, but far more powerful). JAWS can also be used to help you navigate the user interface of Skype and otherwise enhance the Skype experience. To help in this task, a group of volunteers has put together some JAWS scripts for Skype that closely integrate the interaction of JAWS with Skype. These scripts can be found at www.panix.com/~ccn/projects/jfw/skype.php.

The market for accessibility tools for Skype is clearly at the embryonic stage. But there's no doubt that it is set for healthy growth. If you have an interest in accessibility tools for Skype, I suggest you keep an eye out for what's new by searching on the Internet from time to time using a search string such as: Skype (accessibility OR usability).

Operate Skype Using Only Your Voice

Audiomatic is a $10 utility with a free add-on for Skype that enables you to control Skype using only your voice. As an added bonus, Audiomatic will control many of your other applications as well!

After downloading and installing Audiomatic (www.wiseriddles.com) and its free Skype add-on, you will be asked by Skype to give it permission to use Skype (see Chapter 16). Having given Audiomatic permission to use Skype, you can start creating your own Audiomatic *macros* that can then be used to control Skype using only your voice.

def•i•ni•tion

A **macro** is a sequence of instructions that emulate some actions that you would otherwise have to carry out manually. Macros are stored in files and are often used to automate repetitive tasks and to extend the functionality of the applications you use, without the need to modify the application itself. In the case of Audiomatic, macros are used to guide the actions of Audiomatic when it controls other applications using your voice.

To get started with Audiomatic, you should first create a new folder to hold all your Skype macros. To create a new folder within Audiomatic, click on **My Macro Collection** and then go to **Audiomatic: File: New: Folder,** and give it the name **Skype,** as shown in the following figure.

Next, click on the **Skype** folder to select it, and then go to **Audiomatic: File: New: Macro ...** and in the window that appears (see the next figure) give the macro a name; for example, "Call Dad" (enter it without the quotes).

Click on the **Next** button to advance to the next screen. From the **This macro** pull-down menu, choose the option **Interacts With Skype.** In the box labeled **When this macro is run I want to,** there will appear two pull-down menus, as shown in the following figure. The first pull-down menu provides two options: Place A Call To, and Send An Instant Message To. The second pull-down menu contains all your Skype contacts. Continuing with the "Call Dad" macro example, select **Place a call to** from the first pull-down menu, and then select an entry from your contacts list in the second pull-down menu (in this example it would be the Skype user name or telephone number of your father); then click on the **Next** button.

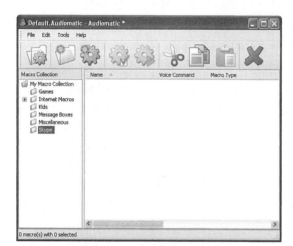

The *main window of Audiomatic, with a Skype macros folder.*

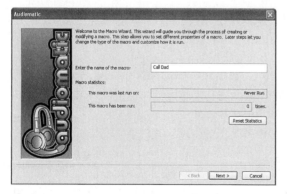

Naming a new Audiomatic macro.

Configure the Audiomatic macro so that it will work with Skype.

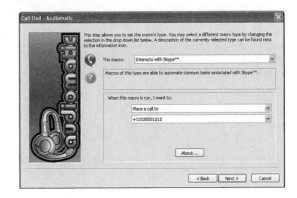

Caution!

In Audiomatic, when creating a Skype macro, if you choose **Send an instant message to** as the action of a Skype macro, the pull-down menu of your contact list retains regular phone numbers for use with SkypeOut as well as Skype user names. However, instant messaging, or chat as it's known in Skype, only works with Skype user names, so creating an Audiomatic macro for Skype chat that uses a regular phone number simply won't work. You can only create chat macros that use Skype user names.

The next screen to appear allows you to select when a macro can be run. The default is **Always available,** so to make your macro always available, leave the default alone and click on the **Next** button. This will take you to the screen shown in the following figure. Put a check mark opposite **Enable Voice Commands for this macro,** and in the text box below the label **Enter the voice command that will activate this macro** put "Call Dad" (enter it without the quotes). Then click the **Next** button and in the next screen that appears click on the **Finish** button.

Congratulations! You've created your first voice-activated macro for Skype.

Now, whenever Audiomatic is ready and you say "Call Dad," Skype will place a call to the number or Skype user name you associated with this macro. You don't need a command to hang up a call, as that will be done automatically when the other person hangs up.

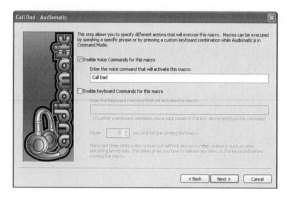

Give the Audiomatic macro a voice command to run it.

If you want a totally hands-free Skype experience, you will also need to have Skype automatically answer incoming calls (**Skype: Tools: Options ...** and in the window that appears, click on the **Advanced** category, and then put a check mark opposite **Automatically answer incoming calls**). With that done, and with Audiomatic running, you're all set to use Skype using only your voice. Look Ma, no hands!

Using the procedure outlined above, you can create any number of Audiomatic macros to place calls to people in your contacts list (Call Dad, Call Mom, Call Pete, and so on).

You can also create Audiomatic macros that will start Skype chat sessions with Skype users in your contacts list. But in that case, the commands will simply open a chat window and establish a chat session between you and the other person. To carry out a chat session using only your voice you will have to use additional software, such as Microsoft Speech (**Start: Control Panel** and in the window that appears, double-click on **Speech**), that can be used to input chat text by means of your voice and that will also read incoming chat messages out loud.

Experiment and have fun with these technologies, but be aware that speech and voice recognition software is still in its early stages, and you should be prepared for something less-than-perfect, and perhaps even some degree of vexation. But stick with it, because, with practice, you might be pleasantly surprised by the results!

The Least You Need to Know

◆ The usability and accessibility of the Skype softphone continue to improve with each new release.

◆ There are some simple things you can do to fill some of the gaps that remain in Skype's accessibility.

◆ By switching on the accessibility features of Windows, you will also improve the accessibility of Skype.

◆ Accessibility tools and add-ons for Skype are slowly making an appearance, and the trend is encouraging.

◆ Using a $10 add-on for Skype you can configure Skype to make and receive calls using only your voice.

Chapter

22

Tweaking Skype's User Interface

In This Chapter

♦ Assigning a shortcut key sequence to open Skype

♦ Using Skype hotkeys

♦ Accelerating Skype using your keyboard

♦ Making calls and chat from the command line

♦ Adding Skype fast-dial shortcuts for your menu or desktop

♦ Displaying the technical details of a call

This chapter is something of a grab bag of tips and tricks for getting the most out of Skype, in most cases by driving the Skype user interface in some novel way.

For example, if you're proficient with the keyboard, driving Skype using only keystrokes can be easier, and faster, than using the mouse. Not only does using the keyboard for some tasks in Skype make sense from a productivity point of view, it also makes sense from an accessibility point of view.

Bottom line: the tweaks you will find in this chapter are useful to everyone, but especially to those people with physical impediments.

Shortcut Keys and Hotkeys

A number of the tweaks in this chapter rely on *shortcut keys*, otherwise known as *hotkeys*, and key sequences. When creating and using shortcut keys, you will typically use these keys and actions:

- **Ctrl:** Press the key marked Control (or Ctrl) on your keyboard.

- **Shift:** Press the key marked Shift on your keyboard.

- **Alt:** Press the key marked Alt on your keyboard.

- **Enter:** Press the key marked Enter on your keyboard.

- **PgDn:** Press the key marked PgDn on your keyboard.

- **PgUp:** Press the key marked PgUp on your keyboard.

- **DnArrow:** Press the key marked with a downward pointing arrow on your keyboard.

- **UpArrow:** Press the key marked with an upward pointing arrow on your keyboard.

def•i•ni•tion

Shortcut keys are short key sequences, typically of two or three key presses, that carry out some action within Windows or for an application. In Skype, shortcut keys are known as **hotkeys**. They are a convenient means of accelerating some tasks, allowing you to do in a few keystrokes what you would otherwise have to do by navigating and clicking with the mouse.

- **+:** This means you must hold down the keys together. So, for example, Ctrl+Alt+A means you must hold down the Ctrl key, the Alt key, and the A key at the same time.

- **[]:** Groups of key sequences enclosed between a [and a] should be carried out in sequence. For example, [Ctrl+Alt+A][Alt+F] means that you should hold down the Ctrl, Alt, and A keys; release them; and then hold down the Alt and F keys.

Assign a Shortcut Key Sequence to Open Skype

By assigning a shortcut key sequence to the Skype icon on your desktop, you can quickly open Skype regardless of what other applications are running.

Here's how you can assign the shortcut Ctrl+Shift+S to the Skype icon on your desktop.

Right-click on the Skype icon on your desktop, and from the popup menu that appears, choose **Properties.**
This opens the Skype Properties window, as shown in the next figure.
Click on the tab labeled **Shortcut** and

> ### Something Worth Knowing
>
> By assigning a shortcut key sequence to the Skype desktop icon, you can pop open Skype even if it isn't currently running. In that case, the key sequence will first run, and then open, Skype.

then click on the text box opposite **Shortcut key.** Now hold down the **Ctrl,** the **Shift,** and the **S** keys simultaneously, and Ctrl+Shift+S should appear in the text box. Click on the **OK** button and you're done.

The Skype desktop icon properties window.

> **Caution!**
>
> If, when assigning a shortcut key sequence to the Skype desktop icon, you hold down Ctrl+Shift+S keys and nothing happens (that is, the shortcut key is shown as "None"); then the most likely explanation is that the key sequence Ctrl+Shift+S is already being used for something else. Try another key sequence instead, such as Ctrl+Alt+S. Keep trying until you find a key sequence that isn't in use anywhere else in Windows.

Now, whenever you use the shortcut key sequence Ctrl+Shift+S, Skype should pop open on your screen and have input focus; that is, the Skype softphone window should come to the foreground of your desktop with its title bar highlighted and be ready to take input from the keyboard.

Skype Hotkeys

You can control certain functions of the Skype softphone using hotkeys assigned by Skype. To enable hotkeys in Skype, go to **Skype: Tools: Options ...** and in the window that appears click on the **Hotkeys** category, then put a check mark opposite **Enable global hotkeys,** as shown in the following figure. To enable specific hotkeys, click on a hotkey in the list to select it, and then click on the **Enable Hotkey** button; this will change the status of the hotkey to Enabled and toggle the text on the button you have just pressed to Disable Hotkey. Clicking again on the Disable Hotkey button will do the reverse and disable the hotkey.

The Skype hotkeys option window.

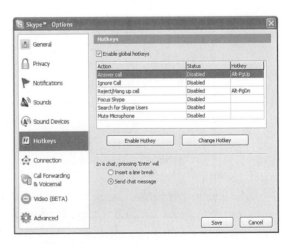

With Skype hotkeys enabled, and individual hotkeys also enabled, the keystroke **Alt+PgUp** will pick up a call, and **Alt+PgDn** will hang up a call. You can assign your own hotkeys to any, or all, of the following actions in Skype:

◆ Answer a call

◆ Ignore a call

◆ Reject or hang up a call

◆ Focus Skype (that is, give the Skype softphone window, mouse, and keyboard focus for input)

◆ Search for Skype users (this opens the Search for Skype Users window)

◆ Mute microphone (this allows you to toggle your microphone on and off)

To set up or change the hotkeys assigned to one of the actions in the previous list, open the hotkeys options window (as described previously). Click on one of the available actions—for example, **Mute Microphone.** Next, click on the **Change Hotkey** button, which opens the window shown in the following figure. Enter the hotkey sequence you want, say Ctrl+Shift+M, and then click the **OK** button. Finally, click on the **Enable Hotkey** button to enable the new hotkey assignment, and then click on the **Save** button.

Assigning a hotkey to a Skype action.

Now, whenever you enter the hotkey you have set up, Skype will carry out the corresponding action. In the example we've been using, this means that when you enter Ctrl+Shift+M you will mute your microphone; if you enter Ctrl+Shift+M again you will unmute your microphone; that is, you can toggle between switching your microphone off and on by using the key sequence Ctrl+Shift+M.

Accelerate Skype Using Your Keyboard

Pretty much anything you can do with your mouse, you can also do by entering key sequences from the keyboard. This can often be much faster than using your mouse, but the catch is that you must remember the correct key sequences! However, for many—especially those who must use the keyboard rather than the mouse, such as the blind—it's an investment worth making.

If you want to use the keyboard to accelerate Skype, your efforts will be made a lot easier if you set up a shortcut key sequence to open and provide input focus for the Skype softphone. Otherwise, your keystrokes might be sent to the wrong application. I described how to set up this shortcut key sequence earlier in this chapter, with instructions for assigning the shortcut key sequence Ctrl+Shift+S to the Skype icon on your desktop. What follows assumes you have set up this shortcut key sequence for Skype. With this in place, entering Ctrl+Shift+S at the keyboard will open Skype and give it focus wherever it might be lurking!

Now you're ready to start driving Skype from the keyboard. To help you more easily understand the concepts and mechanics behind using your keyboard to accelerate Skype, here are some examples for you to try:

- **[Ctrl+Shift+S][Alt+F][Enter][A]:** Will change your online status to Away.

- **[Ctrl+Shift+S][Alt+T][R][Enter]:** Will reopen your most recent chat session.

- **[Ctrl+Shift+S][Alt+T][L]:** Will clear your history.

- **[Ctrl+Shift+S][Ctrl+Tab]** (repeat [Ctrl+Tab] to cycle through tabs): Will cycle through the tabs (Contacts, Dial, and History) in the Skype softphone.

- **[Ctrl+Shift+S][Alt+O][DnArrow][Enter]:** Will open the Search for Skype Users window.

Don't be afraid to experiment with this method of driving Skype. It often takes a little trial and error (and sometimes a notepad and pencil to help you remember the steps) to find the right key sequence for what you want to do. Of course, this method works best when key sequences are short and easy to remember.

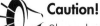

Caution!

Skype hotkeys might conflict with those you've setup for other applications, whereas driving Skype's main menu and user interface from the keyboard always works as advertised—provided, of course, you first give the Skype softphone window input focus and remember to use the correct key sequence!

Make Calls and Chat from the Command Line

You can make calls and start chat sessions directly from the command line; that is, by issuing commands from the keyboard directly to Windows rather than by using the Skype softphone window. The simplest way to run a command on Windows is to go to **Start: Run ...** and in the window that appears (see the following figure), type a command, and then either hit the enter key or click on the **OK** button.

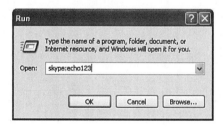

Making a call from the command line.

To make a call from the command line, enter a command such as **skype: echo123** (or skype:+1-203-555-1212 for a SkypeOut call), where echo123 is the name of a Skype user. This will open Skype and start a call to the named Skype user (or telephone number).

Similarly, to start a chat session from the command line, enter a command

Something Worth Knowing

Making calls and chats from the command line also has the advantage that if Skype is not running, Skype will first open, and then the call will be made or the chat session started.

such as skype:echo123?chat, where echo123 is the name of a Skype user. This will open a Skype chat session window with the named Skype user.

Skype Fast-Dial Shortcuts for Your Menu or Desktop

Making a Skype call typically goes something like this: double-click the **Skype** icon in the system tray or the Skype shortcut on the desktop, navigate to the **Contacts** tab, find the entry for the contact you want to call, and then double-click on it. For people you call frequently, these maneuvers can eat up precious time. Wouldn't it be nice to call people with a double-click of your mouse? The good news is that you can—and it's easy!

Menu Shortcuts

To create a menu shortcut for Skype, navigate to Skype on the Windows menu (**Start: All Programs: Skype**), and right-click on the **Skype** icon. A pop-up menu will appear, from which you should choose **Create Shortcut.** This creates a new shortcut under the Skype menu, as shown in the following figure.

A new Skype menu shortcut:
"Skype(2)".

Next, right-click on the new shortcut ("**Skype(2)**") and choose **Properties.** This opens a window in which you can edit the properties of the shortcut you have just created, as shown in the next figure. With the shortcut properties window open, follow these steps:

1. Go to the **General** tab and rename the shortcut. In the example shown in the following figure, the shortcut has been named echo123. For convenience, name the shortcut to match the name of the person you intend to call by using it.

2. Go to the **Shortcut** tab and in the **Target** box, enter the command: **explorer skype:echo123** for a call shortcut, or **explorer skype:echo123?chat** for a chat shortcut.

3. If you want to associate a shortcut key sequence with the menu shortcut, click on the **Shortcut key** text box and enter the key sequence you want to use for this shortcut using your keyboard; for example, Ctrl+Shift+E.

4. Click the OK button and you're done.

The properties window for a shortcut.

Something to Try

If you have previously assigned a speed-dial number (see Chapter 10) to the contact you are associating with a shortcut, it makes sense to use this as part of the shortcut key sequence. For example, if the person for whom you have set up a shortcut has the speed-dial code 4 in your contacts list, use 4 as part of the shortcut key sequence, say Ctrl+Shift+4. Pressing Ctrl+Shift+4 will open Skype and start a call with that person. Unfortunately, while Skype speed-dial codes go from 0 to 99, you can only use the digits 0 through 9 as part of a shortcut key sequence.

To use the menu shortcut you have made, either navigate to the shortcut using your mouse (**Start: All Programs: Skype: echo123**), or simply use the shortcut key sequence Ctrl+Shift+E. Using this method, your most frequent calls and chats are only a few mouse clicks or keystrokes away.

Desktop Shortcuts

To create a fast-dial shortcut for Skype on your desktop, you have a couple of choices.

First, if you already have fast-dial menu shortcuts (see the preceding section), you can simply right-click on any of them, and from the popup menu that appears you can choose **Send To: Desktop.** This creates an icon shortcut on your desktop with the same functionality.

Second, you can simply right-click anywhere on an unoccupied area of your desktop. From the popup menu that appears, select **New: Shortcut.** This creates an empty shortcut on your desktop and also opens the **Create Shortcut Wizard** window, as shown in the following figure. In the wizard, follow these steps:

1. In the text box below **Type the location of the item,** enter the command: **explorer skype:echo123** for a call shortcut, or **explorer skype:echo123?chat** for a chat shortcut. Then click on the **Next** button.

2. In the text box below **Type a name for this shortcut,** enter a name for the shortcut, say echo123. Then click on the **Finish** button. This creates a new desktop shortcut.

3. If you want to change any of the properties of the newly created shortcut, right-click on it and from the popup menu that appears choose Properties. This opens the properties window for the shortcut. Make any necessary changes and click on the OK button.

Note that a careful naming convention for your Skype fast-dial desktop shortcuts will pay dividends when it comes to organizing your desktop. This is particularly true if you have more than a few such shortcuts. Using a short prefix, such as "SkypeCall–" or "SkypeChat–" (without the quotes), for shortcut names will mean that when you organize your desktop icons by name, all your fast-dial shortcuts will group together nicely as a block. This will make finding and calling your most frequent Skype contacts quicker and easier still.

Double-clicking on a fast-dial desktop shortcut will start a call (or chat) to whomever you have entered as the target of the shortcut. Skype calls and chats are now just a double-click away!

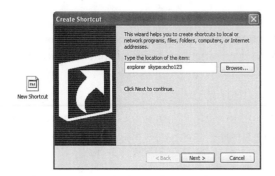

Creating a fast-dial desktop shortcut for Skype using the shortcut wizard.

Display the Technical Details of a Call

It can sometimes be useful to look under the hood of Skype, so to speak. By switching on the technical details for a call feature, you can find out a wealth of technical details for that call simply by making your mouse pointer hover over the picture of someone who is participating in the call, as shown in the following figure.

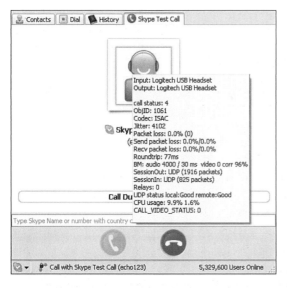

Displaying the technical details for a call.

To switch on the display of technical details for calls, go to **Skype: Tools: Options …** and in the window that appears click on the **Advanced** category, and then put a check mark opposite **Display technical call info** and click the **Save** button. Thereafter, whenever you hover your mouse pointer over the picture of someone you are in a call with, a popup will appear with the technical details of the call and Internet connection between you and that person. This information can be particularly useful when troubleshooting a problem with Skype (see part 5 of this book for tips and techniques on how to troubleshoot Skype).

The Least You Need to Know

◆ Assign a shortcut key sequence to Skype so that you can open and display the Skype softphone with nothing more than a few key presses.

◆ Skype hotkeys can simplify and speed up some actions in Skype.

◆ You can drive Skype more quickly with the keyboard than with the mouse for most actions.

◆ You can make calls and start chat sessions from the command line.

◆ Fast-dial menu and desktop icon shortcuts can put your most frequent contacts just a shortcut key sequence or double-click away.

◆ By switching on technical details for Skype calls, you can peek under the hood of Skype for troubleshooting purposes.

Appendix A

Glossary

802.11   A wireless networking standard for WiFi. Often used as a synonym for WiFi.

access point   A wireless network router configured in such a way that multiple PCs can connect to it at the same time.

add-on   An application that runs independently of Skype, but augments what you can do with Skype. Also known as a plug-in or add-in, such applications typically use Skype's API.

ADSL   Short for asynchronous digital subscriber line, a common type of DSL service in which the data rates for sending and receiving are different.

antivirus program   A program, also known as a virus scanner, that scans files for computer viruses and alerts you to their presence. An antivirus program is one of the most important programs you can have on your PC.

API   Short for application programmer's interface. Skype's functionality can be augmented by applications that are separate from Skype but use its services through an API. The Skype API defines the commands that can be passed to and between Skype and another application and the responses to those commands.

assistive technology Hardware and software that enable people with impaired abilities to make easier and better use of something.

audio input This is an input socket on your PC that allows you to connect an audio-input device, such as a microphone, to your computer. It is usually a 3.5mm-diameter socket that is clearly marked for this purpose. Note that some microphones connect to your PC through the USB port rather than a 3.5mm-jack plug.

audio output This is an output socket on your PC that provides audio output to some sort of speaker, and is usually a 3.5mm-diameter socket that is clearly marked for this purpose. Note that some speakers—typically in the form of a USB headset or USB handset— connect to your PC through the USB port rather than a 3.5mm-jack plug.

avatar Is an abstract, graphical depiction of a person.

bandwidth A measure of how much data per unit of time can be transferred over your Internet connection. Bandwidth is typically measured in kilobits per second (Kbps) or megabits per second (Mbps), where 1 Kbps = 1,000 bits of data per second and 1 Mbps = 1,000 Kbps. The bandwidth for an Internet connection when sending data is often not the same as that when receiving data; a situation commonly referred to as an *asymmetric Internet connection.*

Bluetooth A wireless technology that enables electronic devices to network together at short distances, typically 30 feet or less. Bluetooth headsets and handsets give Skype users some mobility while making and receiving calls.

broadband A term describing a type of Internet connection. A broadband Internet connection is significantly faster, in terms of the amount of data it can transfer in a given interval of time, than a dial-up Internet connection. Skype is best used with a broadband Internet connection, such as that provided by DSL or cable Internet.

bug A catch-all term for anything that goes wrong with a computer program.

cable Internet A type of broadband Internet connection provided by a cable company.

call rounding   A feature of SkypeOut that rounds phone calls to the next whole minute. The first four seconds of a Skype call are free, but after that, you pay in increments of whole minutes.

DSL   Short for digital subscriber line, a type of broadband Internet connection provided over your telephone line.

emoticons   Icons representing emotions, such as happiness, sadness, confusion, and the like.

encryption   The technique of taking data and garbling it in such a way that no one else other than the owner or intended recipient of the data can understand it.

firewall   A gatekeeper for your Internet connection. It blocks malicious network traffic from reaching your computer, thereby providing a more secure Internet experience. The functions of a firewall can be provided by either a piece of networking hardware, or simply software on your PC.

hotkeys   See *shortcut keys.*

hotspot   A location where wireless connectivity to the Internet is available, although not necessarily available to everyone or for free. Hotspots are typically no more than a few hundred feet in size.

HTML   Hypertext markup language. This is a fairly simple text-based language that describes to web browsers how a web page should be displayed.

Kbps   See *bandwidth.*

knowledgebase   A database of information about a particular topic, structured in a searchable format.

latency   The time delay between when you start talking on your end of a call and when the recipient on the other end hears your words. All network connections have some latency. A latency in excess of 500 milliseconds (half a second) makes natural conversation almost impossible, as you and the call recipient will most likely start talking over one another.

macro   A sequence of instructions to a piece of software that emulates some actions that you would otherwise have to carry out manually.

malware Any software that damages or degrades the performance of your PC, or steals or damages your data in any way. See also *spyware*.

Mbps See *bandwidth*.

node A computer connected to a network. For the purposes of this book, a node is a PC running the Skype softphone that is connected to the Skype P2P network via the Internet.

P2P Short for peer-to-peer. A network configuration in which each computer on the network runs the same software and is treated equally; that is, every computer is connected to one or more peers.

PC Short for personal computer. Specifically, for the purposes of this book, a computer that runs Microsoft Windows.

Personalise Skype A fee-based Skype service that allows you to buy sounds, ringtones, and pictures that jazz up your online image and enhance your Skype experience.

POTS Short for Plain Old Telephone System, a term referring to the services provided by your regular telephone system.

proxy A computer that is placed between your PC and the Internet, and whose purpose is to speed up, make more secure, and otherwise improve your interaction with the Internet.

router A network device to which you can connect one or more PCs so that they form a network, and which provides common access to the Internet for all PCs on that network. Some routers use wires to connect to PCs, others use wireless networking; some routers have the ability to use either, or both.

shortcut keys Short key sequences, typically of two or three key presses, that carry out some action within Windows or for an application. In Skype, shortcut keys are known as hotkeys.

Skype Control Panel A Skype service that enables a single person to administer Skype services (SkypeIn, SkypeOut, voicemail, and so on) for a group of people.

SkypeIn This fee-based service enables you to receive calls from a regular or mobile phone by giving your Skype account a dial-in number. Dial-in numbers are available for many different countries. Each dial-in number costs $36 for one year or $12 for three months.

SkypeOut   This fee-based service allows you to call regular and mobile phone numbers worldwide. Per minute rates are as low as $0.021.

Skype Zones   For people who are on the move and travel around a lot, this fee-based service gives you access to more than 18,000 Skype-friendly wireless access points worldwide.

softphone   A softphone is a computer program that emulates the functions of a telephone, enabling users to make calls over the Internet.

sound-in device   See *audio input.*

sound-out device   See *audio output.*

spam   Any message sent via voice, chat, or e-mail that is both unsolicited, and unwanted by the recipient. Voice spam is also known as spam over Internet telephony, or SPIT.

spyware   Unwanted software that—unknown to you—monitors what you do on your PC and reports such activity to someone else.

supernode   A Skype node that has been designated to take on some of the management functions for Skype's P2P network. Skype nodes can become supernodes, and vice versa. Whether your PC becomes a supernode on Skype's network is something over which you have no explicit control.

TCP/IP and **UDP**   Two of the networking protocols that underlie the Internet. You will come across these terms when configuring and troubleshooting Skype's network settings. All you need to know is that Skype can use both TCP/IP and UDP for transferring data, and that both use one or more "port numbers." Port numbers, which are numbered from 0 to 65535, are analogous to regular phone sockets, in the sense that in order to use a telephone handset to connect to a phone line, you must plug it into a specific socket. In a similar fashion, Skype must use specific TCP/IP and UDP ports to establish a telephone connection.

USB   Short for universal serial bus; a common method by which to attach peripherals to your PC. In particular, for Skype, it is a common method by which to attach a USB headset or USB handset to your PC.

USB handset   A microphone and speaker built into a telephone handset that connects to your PC through a USB port. Such handsets are commonly used as the audio input and output device for Skype, and have the advantage of sounding and feeling like a regular telephone handset.

USB headset   An earphone headset and microphone (the latter usually supported on a boom in front of the mouth of the wearer), built as a single unit; the headset connects to your PC through a USB port. Commonly used as the audio input and output device for Skype.

USB hub   A device that expands the number of available USB ports for your PC.

vCard   A standard format by which to exchange personal data, specifically the sorts of data you might find on someone's business card. Skype's archive utility (backup and restore) exports and imports your contacts using the vCard format.

virus   An unwanted self-replicating and self-propagating computer program that can do great harm to your PC.

webcam   A video camera that attaches to your PC, usually through a USB port, and provides real-time video input for applications that run on your PC.

WiFi   Short for Wireless Fidelity, a wireless network technology that has a limited range. WiFi can support broadband Internet connections and is a great way to provide some mobility to Skype users. See also *802.11.*

Resources

This appendix lists additional resources on Skype.

Skype User Forum

By far the most extensive and active Skype resource is its own forum, which can be found at forum.skype.com. Anyone can use the forum, and it's free! The forum is divided into areas of interest and is frequented by some of the most knowledgeable Skype users around. A question or problem posted here will most likely get you a response within 24 hours.

Other Web Resources

Here are a handful of Skype support, commentary, and community websites you may find useful.

Skype Support

Skype's searchable online help knowledgebase can be found on the web at support.skype.com. There you can search for answers to a lot of common, and some not-so-common, problems. You can also post your own support requests.

The Skype Journal

The Skype Journal is a popular online magazine which can be found at www.skypejournal.com. It's a great source of information for everything that's going on in the Skype worldwide community. Most of its content is of interest to new and intermediate Skype users, although some of its articles are quite technical.

Elpis's Skype Power User

Elpis's Skype Power User magazine (www.elpispublishing.com) is an electronic magazine that seeks to educate the Skype user about useful and nifty tools and utilities to use. The mantra at Elpis is knowledge + tools = empowerment, which is a convenient way to express the idea that knowledge can only be unleashed if you have the tools to do so. In the interest of full disclosure, you should be aware that the author of the book you're holding in your hands is also the publisher of this online magazine.

Jyve

Jyve, www.jyve.com, is an online community of Skype users. Jyve enables its community members to participate in any of its user groups that span a wide range of interests. Groups and their members are searchable. So, for example, if you want to find a speaking partner to practice a foreign language, this is one place to go.

Books

What follows is a list of books specifically on Skype. None come with any particular recommendation from the publisher or author of this book, but they are alternative sources of information on Skype. The following Skype books were in print as this book went to press:

Skype Hacks: Tips & Tools for Cheap, Fun, Innovative Phone Service, by Andrew Sheppard (yes, that's me again; O'Reilly Publishing, 2005).

Skype Me!: From Single User to Small Enterprise and Beyond, by Bill Campbell. (Syngress, 2005).

Both books are targeted at the intermediate to advanced Skype user.

Appendix C

Skype Add-On Software and Hardware

Skype has an Application Programmer's Interface, or API, that allows applications that are separate from Skype to use Skype's functionality and services. There is a burgeoning market for third-party add-ons that use Skype's API to augment what you can do with Skype. Many are free, but some you have to purchase.

Some add-ons (also sometimes referred to as plug-ins, or add-ins) use only software, whereas others use a combination of software and hardware. Both types of add-ons are covered in this appendix.

Before delving into this topic any further, a word of caution is in order: like any software you download from the Internet, Skype add-ons pose a security risk. Before downloading an add-on, you should ask yourself a few questions. Do I need it? Who wrote it? Can I be sure that the add-on won't do something bad? Downloading files and programs is one way in which your PC can get infected with a virus. So, at the very least, anything you download from the Internet should be passed through a virus scanner.

Software Add-Ons

Software add-ons for Skype are applications that you can download, install, and run on your PC alongside Skype.

When you run a Skype add-on for the first time (and if you subsequently update it), as a security measure, Skype will pop up a window asking permission for that add-on to use Skype, as shown in the following figure. Other applications and add-ons will not be able to use Skype unless you give them permission.

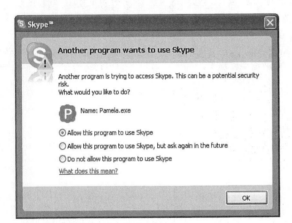

Giving permission for an add-on to use Skype.

Even though Skype is a young product, even by Internet standards, there are already a bewildering assortment of add-ons available. One of the best places to visit to find add-ons for Skype is the Skype Extras Gallery, at share.skype.com/directory. Here you will find lots of add-ons grouped by category and, most importantly, many have been rated by Skype users.

Hardware Add-Ons

First, let me define what I mean by a Skype hardware add-on.

Devices that plug into your PC and provide basic audio input and output or video, such as a USB headset or webcam, do not qualify as Skype hardware add-ons because they don't use the Skype API. Such devices are simply PC peripherals that Skype happens to use.

A Skype hardware add-on is an item of hardware that plugs into your PC, usually via a USB port, and which uses the Skype API to make that hardware work cooperatively with Skype. That is, you must install both the hardware and some software for a Skype hardware add-on to work. As is also the case for software-only add-ons, you will have to give the hardware add-on software component permission to use Skype for the whole thing to work.

One common type of Skype hardware add-on that you will likely come across is an analog telephone adapter (ATA), which enables you to use a regular or cordless phone handset with Skype. A typical ATA device is shown in the following figure.

A typical analog telephone adapter (ATA).

Using an ATA, and by installing the software that operates it, you can control Skype and make Skype calls from a regular or cordless handset. This way of making calls over the Internet has three distinct advantages. First, you can use a regular or cordless phone handset, which is something you probably already have. Second, you have the familiarity of a device that you, and others, are already comfortable using. Third, if you plug in a cordless phone base station into the ATA, you can use any cordless handset from anywhere throughout your home or office to make and receive Skype calls (as well as regular calls, if you are a SkypeOut and SkypeIn subscriber). A Skype compatible ATA will set you back around $40.

Appendix D

A Guide to Skype's Graphical User Interface Elements

Nowhere on the Skype website or its online tutorials are the graphical elements of its user interface named or described. However, for someone starting out with Skype, knowing what things are named and what they do is essential. The purpose of this appendix is to fill this particular knowledge gap. While not truly comprehensive, this appendix does describe the most important graphical elements of Skype.

At the very least, if you know the correct names of these elements you will know what terms to use when you seek help using Skype's knowledgebase, user forums, or when submitting a support request. And in the case of the latter, the person who receives it will know what you're talking about!

Graphical Element	Name	Purpose
	Call button	Click on this button to start a call
	Hang-up button	Click on this button to end a call.
	Add Contact button	Click on this button to add a person to your contact list.
	Search for Skype Users button	Click on this button to open a window that will allow you to search for Skype users.
	Conference Call button	Click on this button to set up a conference call, or add contacts to an existing call.
	Chat button	Click on this button to start a chat session.
	Send File button	Click on this button to send a file to a single Skype user.
	Send File to Many button	Click on this button to send a file to multiple Skype users.
	View Profile button	Click on this button to pop up a window that gives summary information about your Skype profile.

	Online Status is Offline	Wherever you see this icon, it means that the Skype user is offline.
	Online Status is Online	Wherever you see this icon, it means that the Skype user is online.
	Online Status is Skype Me	Wherever you see this icon, it means that the Skype user is in Skype Me mode.
	Online Status is Away	Wherever you see this icon, it means that the Skype user is away from Skype.
	Online Status is Not Available	Wherever you see this icon, it means that the Skype user is not available.
	Online Status is Do Not Disturb	Wherever you see this icon, it means that the Skype user does not want to be disturbed.
	Online Status is Invisible	Wherever you see this icon, it means that the Skype user has chosen not to make his or her online status visible to others. However, when online, the privacy settings for that Skype user remain in effect.
	Online Status is Call Forwarding	Wherever you see this icon, it means that the Skype user is offline and has set up Skype to forward calls.
	All Events icon	This icon is used to denote all events.
	All Calls icon	This icon is used to denote all calls, both incoming and outgoing.

	Missed Call icon	This icon is used to denote a missed call, or missed calls.
	Incoming Call icon	This icon is used to denote an incoming call, or incoming calls.
	Outgoing Call icon	This icon is used to denote an outgoing call, or outgoing calls.
	Voicemail icon	This icon is used to denote events and actions related to Skype voicemail.
	File Transfer icon	This icon is used to denote a file transfer, or group of file transfers.
	Chat icon	This icon is used to represent events and actions related to chat.
	Play button	Click on this button to play a voicemail or a sound.
	Stop Playing button	Click on this button to stop playing a voicemail or a sound.
	Record button	Click on this button to record sounds.
	Reset button	Click on this button to reset a setting.
	Trash (Delete) button	Click on this button to delete an item.

🎤	Mute button	Click on this button to mute your microphone. Click it again to unmute.
⏸	Hold-Call button	Click once to put a call on hold. Click again to take a call off hold.
📞	Call Phones button	Click to sign up for SkypeOut. If you are already a SkypeOut subscriber, clicking this button takes you to the Dial tab.
🚩	Event icon	Denotes that some event has occurred.
📹	Video icon	Denotes that a Skype user has video.
📹 Start My Video	Start Video button	Click to start video.
📹 Stop My Video	Stop Video button	Click to stop video.

Index

Symbols

+ shortcut key, 204
.wav extension, 109
[] shortcut key, 204
411 directory services, workarounds, 190
911 emergency services, workarounds
"911-like" services, 189
mobile phone backups, 189-190
service by default, 188-189

A

accessibility features, 195
Audiomatic, 198-201
speech recognition programs,
197-198
Windows, 196-197
Accessibility Wizard, 196
access points, Skype Zones, 68-70
activation
SkypeIn, 55
Skype Voicemail, 58-59
Add a Contact window, 91
Add Selected Contact button, 90
add-ons, 125
hardware, 127
locating, 126-127
security management, 155-156
software, 128-129
adding contacts, 90-92
Address bar (graphical interface),
32
Advanced category, configuration
of options, 101-102

adware, as privacy threat, 136
Alt shortcut key, 204
alternatives to Skype, 12-13, 76-77
antifraud restrictions, SkypeOut,
50
API (Application Programmer's
Interface), 126
Application Programmer's
Interface (API), 126
applications
Skype toolbar for Microsoft
Internet Explorer, 121-122
Skype toolbar for Microsoft
Outlook, 117-119
architecture, 7-8
archiving contacts, 96
assistive technologies, 196
associate sounds, configuring
online persona, 111-112
AsYouGo pricing option (Skype
Zones), 66
audio, troubleshooting, 173
error codes, 178-179
option settings, 173
testing, 174-176
Audio In sound device, 100
Audio Out sound device, 100
Audiomatic, 198-201
authentication, security management, 157-158
auto-sign-in, privacy settings, 145
auto-start, privacy settings, 145
automatic proxy detection, 183
Automatic Updates window, 157
Available Signals link, 70
avatars, 106
Away (online status), 86

B

bandwidth requirements, 42
blocking unwanted contacts, privacy settings, 146

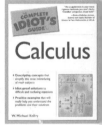